GET UP & GO

A TRAVEL SURVIVAL KIT FOR WOMEN

ÁINE McCARTHY

Attic Press
Dublin

© Aine McCarthy 1992

All rights reserved. Except for brief passages quoted in newspaper, magazine, radio or television reviews, no part of this book may be reproduced in any form or by any means, electronic or mechanical, including photocopying or recording, or by any information storage and retrieval systems without prior permission from the Publishers.

First Published in 1992 by
Attic Press
4 Upper Mount Street
Dublin 2

British Library Cataloguing in Publication Data
McCarthy, Aine
    Get Up and Go: Travel Survival Kit for Women
    I. Title
    910.2

ISBN 1- 85594-037-X

Cover Design & Illustrations: Cathy Dineen
Origination: Attic Press
Printing: The Guernsey Press Co. Ltd

*For my mother, Ida, and father, Con.
Without them I could never have gone anywhere.*

Contents

Preface by Aine McCarthy	7
Why leave your armchair?	9
OK, who's going then?	13
You want to go where?	45
Getting the paperwork out of the way	87
Staying healthy and well	96
A - Z	111
Further reading	151
Contact addresses	153
Index	156

Preface

From Bali to Baltimore, Mongolia to Mexico, Alaska to Austria, the world is opening up, becoming more and more accessible. Travel is now the world's fastest growing industry, and women are travelling further afield than ever before.

To match this trend hundreds of guides, brochures, magazines, films and videos have been produced, showing you where to go, how to get there, and what to see and do. Unfortunately, in most of these travel guides women are almost invisible. No attempt is made to grapple with the specific problems we face in getting going and staying safe.

The woman who wants to travel faces all the usual joys and pleasures, difficulties and dilemmas: choosing where to go, coping with foreign cultures and climates, and so on. Being a woman can bring an extra set of problems, from fending off unwanted attentions and sexual harassment to coping with the various dress and behaviour codes which different societies have decreed appropriate for females.

This book puts women into the travel picture, directing the general sort of information which all travellers require towards the special needs and interests of women. It gives you information which will save you time and trouble before you leave, and keep you safe and happy on your journey through any part of the world. Remember, there are women everywhere, and where they thrive so can you!

The book is divided into sections. The first, **Why leave your armchair?**, looks at the history of women travelling and talks to some women who have discovered the delights of travel. The qualities needed to embark on a journey are independence, curiosity about the larger world, resilience and a healthy selfishness, not qualities which most societies admire or nurture in the female sex. The testimony of these wayfarers should convince you that travel is worth it, and encourage you to get up and go.

How and where you go will be affected by your situation - whether you are a young woman, or older, whether you have children or are pregnant, whether you are able bodied, disabled or retired. The second section, **OK, who's going then?**, looks at the specific problems which different women encounter when they travel, and how those problems can be surmounted. It also looks at how travelling companions, or lack of them, will affect your trip.

The third section, **You want to go where?**, gives information about the regions of the world, Africa, Asia, Europe, Middle and Far East, the Indian Subcontinent, North and South America. Climate, regional customs and traditions pertaining to women are included plus travel tips to help you get the most from a visit to your chosen area.

The fourth section, **Getting the paperwork out of the way** does just that: puts travel documentation in its place and explains how best to handle money on the move.

Staying healthy and well contains important information on health and safety measures for travellers: vaccination, hygiene and medication. The fear of sexual harassment or attack is treated in detail here, along with some hints on self-defence.

The **A-Z** section is a compendium of information on miscellaneous matters: helping you to cope with everything from fear of flying to menstruation, from culture shock to camping. This is followed by useful *check-list*s.

An extensive bibliography of further recommended reading, and a list of travel associations and contact organisations that any travelling woman will find helpful are also included. So what are you waiting for? Get up and go!

Áine McCarthy

Why leave your armchair?

In her book *The Blessings of a Good Thick Skirt* travel writer Mary Russell addresses the question of why women 'soar off into the dawn skies, trudge across deserts, sail into uncharted waters or cling perilously to the peaks of snow-bound mountains'.

Travel can be as simple as two relaxing weeks in the sun. But the question remains, why leave the comfort and safety of our homes? Why stretch our legs beyond the armchair? Why bother?

The reasons, says Russell, are myriad: 'to escape from domesticity or the drudgery of a routine job; to recover from a broken love affair; to experience the thrill of danger; to demonstrate that woman's name is definitely not frailty; to bring the Bible to China; to study plant life or unknown peoples; to delve into the past; to expiate a private guilt; to honour a dead partner; to glorify their country; to find something interesting to write about - or simply to have fun. That some set out with no motive other than to enjoy themselves is clear - and to me this is the best reason of all.'

That we do bother, and in increasing numbers, cannot be denied. Tourism is today's fastest growing industry, with over forty million people travelling for leisure in 1991. Once women have access to money it seems they want to travel.

There is nothing new about women travelling the world, although mass tourism has allowed more women than ever before to get on the move. In the distant past the majority of women who travelled did so for religious reasons, taking the pilgrim route. There are records of a religious sister from Roman Gaul called Egeria who travelled to Jerusalem as early as 383 AD. Later, religious communities enabled women to travel by providing the money and incentive to study at sister convents and religious centres around the world. This tradition of travel within the religious life continued in different forms, with a great upsurge in the eighteenth and nineteenth centuries as missionaries travelled to the other side of the world in order to bring The Word to Africa and Asia. The tradition continues to this day.

Outside the religious life the first women to travel in great numbers came from the aristocracy of Europe. The custom of the 'Grand Tour' began in the seventeenth century and wealthy, upper-class women travelled to major European cities, admiring architecture and culture and writing fascinating diaries of their experiences.

At the end of the nineteenth century new modes of travel, the bicycle, motor car and aeroplane, opened up fresh opportunities for travel. And since the 1950s the cost of such transport has became relatively cheaper. This has led to an explosion in mass tourism, so that today women are taking to the road (or sea or sky) in ever increasing numbers.

Women have decided that travel *is* worth it. A long trip can teach you a tremendous amount about yourself and the world you live in. And a spell in another country is probably the best use you can make of your annual fortnight break, providing you with a complete contrast to your usual routine. Travel allows you to look at yourself from a different perspective. You should return recharged, more appreciative of your daily life, or ready to make necessary changes.

There are as many reasons for deciding to go as there are trips to take. Here meet some women who have told me why they travel and see just how varied their reasons are:

'I am thrilled by the strangeness of currencies and language, and the sameness of communication and body language. When I see how others live it's not only educational, it can be revelatory. Getting some understanding of the forces that drive their lives brings me nearer to understanding myself and my own place and time, my own geography and history.'

Anne, journalist

'I visited Australia last year and that allowed me to say that I've seen parts of all five continents now, which gave me a little thrill. At different times I've travelled alone, with one or more female friends, and with my family. do I go? To see the world, to learn, to escape for a while, but most of all because I enjoy it.'

Dervla, computer operator

'I've done two round-the-world trips and I spend all my spare money on travel. When I was in college I worked away every summer. But it's funny; I never before asked myself why I go. I don't think I have much control over it; it's as if something is pulling me. I'm always glad to get home but within weeks I'm planning the next trip. It keeps me sane and happy. I dread to think what I would have been like if I had lived in an age when women weren't allowed to travel.'

Alma, accountant

'I'm a teacher and so is my partner, so we're really fortunate. Each summer we pack up and head off somewhere with the kids for an extended holiday. Sometimes one of us might teach English while we are away to stretch out the funds; you can usually find somebody in most parts of the world who wants to improve their English. We are lucky that we both love travelling and so do our children, aged four and six now. We find it tremendously exciting. It gets us all through the winter, planning where we will go next. And it's so educational for the children and for ourselves too.'

Mary, teacher

'Every time I go abroad I try to discover something previously unknown to me about the women in the countries I visit. Getting up very early and wandering about before the working day begins, especially in rural towns - this is where I meet and talk to women. Or going along to women's resource centres, social places or libraries in foreign cities is also a way of making contact. Every trip abroad has brought me knowledge about women, knowledge that you just can't get in a travel brochure or guide.

Gráinne, editor

Having calculated the benefits of travel against the risks, women all over the world are jumping out of their armchairs. Why not join them and discover the delights of travel for yourself?

OK, who's going then?

Are you a business woman travelling for work reasons? A mother on a family holiday? A woman with her girlfriend just heading off for some sun? Are you disabled? Pregnant? Retired? All of this has a bearing on how you travel.

Of course the outstanding feature of women's lives today is that they are no longer confined to one role for all of their days. The businesswoman who travels for work may go on family trips for her holidays. The woman who is pregnant one year has the trauma of travelling with, or without, her baby the next. You may be on a two-week sun holiday today but who's to say that you won't make a lone round-the-world trip next year?

This section gives you all the advice you need to help you cope with your individual circumstances, whatever they might be, including advice on travelling alone, with a companion or in groups.

Travelling alone

Travelling alone is an intensely personal experience. Women who go it alone always enthuse about it being *the* way to get to know a place really well.

Lone travel forces you to become more involved in the country you have come to experience. Local people are far more likely to invite you to their homes when you are alone than when travelling with a companion or as part of a crowd. You have nobody to cushion you, nobody with whom to make notes. You have to adapt to your surroundings. While that can be daunting, it is also likely to be immensely rewarding.

You should find that travelling alone gives your self-confidence and self-reliance an enormous boost. Walking into a hotel or bar as a lone woman can be nerve wrecking, but when you're travelling you have to eat, and a bar or restaurant which is a male preserve may well be your only option. At first you may be very nervous, but if you adopt an air of confidence, you are unlikely to come to any harm. Do bring a book or newspaper with you. Once you have coped with the stares a few times you put them into perspective, and that confidence you at first pretended to have becomes real.

Many people remark on the aura of strength and calm which lone women travellers have acquired. It really is a by-product of travelling alone. The realisation that you can cope by yourself with tricky people or situations and that you don't need anybody else in order to survive is tremendously liberating.

So what should you consider before and during your lone travels?

*Firstly, ask yourself if this is something you really want to do? If travelling alone will make you miserable, why bother?
*Learn to like being truly on your own. Pamper yourself. Use the time to think, plan, make decisions. Write down useful thoughts. Keep a diary. Enjoy thinking things through. Take advantage of the new perspective which your travels will give you.
*Take advantage of being alone to mingle with local people and find out as much as possible about your host country.
*Learn to put yourself first. Adopt a confident, decisive approach. Don't accept first refusals. Do a lot of smiling, but speak clearly, decisively and relatively loudly.
*If you get to the stage where you need company and aren't getting it on your travels, make for a youth hostel or tourist area where you are likely to meet other travellers. Or take an organised excursion such as a boat trip; these can be useful for meeting people in a 'controlled' environment.

Travelling with a companion

Making plans with travelling companions can be more fun and it can also make the practicalities of travel easier. When you are tired or apprehensive, it's nice to have someone to turn to. You may also feel safer together than travelling alone.

More important, perhaps, is that companions bring their different outlooks and skills to the trip. Sharing your ideas and emotions can make your experience richer and you may well encourage each other to do things you might otherwise not have tried.

But there are disadvantages too, the main one being that you have to learn to compromise. Constantly accommodating yourself to somebody else's timetable can become quite a strain, no matter how friendly you are.

'If you want to know me, come and live with me' - this old adage applies to any people setting off to travel together. The stresses of travelling will exacerbate irritation. The quirk of personality that you regarded with amused tolerance back home can become unbearable when away together.

On a long trip, particularly, you cannot be sure how the travelling will affect you. Such mundane matters as the time of rising can become an issue. India to one person can be an exotic kaleidoscope of colour and interest, to another a distressing manifestation of poverty and dirt.

However, a lot of conflict can be eliminated beforehand. The time to discover that your companion abhors walking is not when you are half way up a mountain. So if you have decided to go with one or more companions, whether you know them well or not, you must sit down before you go and discuss possible areas of conflict.

*Discuss your travel aims. What does each of you want to get from this trip?
*Discuss your budget in detail and be honest about your attitudes to money. Are you expecting to travel at the same financial level with regard to accommodation, food and so on? Do you have the same amount to spend?
*Establish each other's need for privacy, for time to digest and absorb what's happening. This means being alone from time to time.
*Remember you don't have to do everything together. Discuss this before you go, recognising that there are going to be

times when you will want to be alone. Build in the chance to do different things and meet later at your accommodation, or the next destination.

Travelling in groups

Groups may feel safer, and give you a sense of security that might be particularly attractive if you have never travelled before. Organised tours are good for those who want to hand over the responsibility for organisation and arrangements to somebody else. In some countries, such as Albania, independent travellers are discouraged and one can get in only as part of a group. And for certain types of travel, like mountaineering or arctic travel, it is essential to go in a group.

The disadvantage to group travel is that it tends to be inflexible. Spontaneity and freedom can be lost. And organisational problems can take up much time. Inevitably some people in a group feel excluded from planning and decision-making, and in a mixed group it is not unusual for men to attempt to take over at this level.

Another problem can be that what's going on in the group takes over from what is going on around you, so that your reason for being there in the first place can be lost.

So how do you make group travel a success? Ask yourself the following questions:

*If it is an organised tour has the company considered the specific needs of women travellers, and if it is an all-woman group is your courier a woman?
*How much say do individual group members have in what happens on the tour? Is there time set aside for group members to go exploring on their own?
*What sort of briefing material is provided and how does it present the cultures you will encounter?

You may plan from the outset to be part of a group, or you may find yourself becoming part of a group as you go. But, organised or informal, the secret to successful group travel is to sort out the decision-making process before you go and to build in some way by which members of the group can spend time on their own when you get there.

Travelling for business

Business travel differs from other kinds of travel where you may be choosing to go freely. Business dealings in your company may necessitate travel; the combination of your boss, your company and market forces may be making the decision.

Before making any business travel arrangements check any special requirements your host country may have. Saudi Arabia, for example, will permit a woman to enter only on business and will require an appropriate letter from your company on business letterhead paper.

Also check the availability of your business contacts before making any travel plans. Make sure you are not coinciding with any national holidays. Be particularly careful in the summer. France shuts down for August, for example, as do parts of Scandinavia for the month of July. Plan your visit so that it is not too close to the start or the end of a holiday period.

Organise a schedule for yourself. Get a street map of the area you will visit. Set up lunch or dinner appointments as your anchor meetings and work out from these. Do not plan too tightly; be realistic about time and allow for overrun.

Before your visit send details of your business, along with any samples or brochures which might help your business contacts become familiar with your company and products. This will allow them to prepare queries and requests in advance and lead to more productive meetings.

Before you start your business activities in a foreign place, allow yourself some time to get a feel of the place. Read papers and watch TV; such activities may help improve your understanding and business decisions. Do allow time between appointments for market research. Look at the competition and/or acquire trade publications or statistical information which might be important to your strategy for that market.

You should ensure that you enjoy your trip as well as getting your work done. Do some groundwork on the country before you go. Find out about sights and events that you would like. Fit in some time for yourself. Play the tourist; you'll work better as a result.

Each time you visit an area or country for which you have responsibility you should increase your knowledge of its economic, political and social climate. Ask questions of your clients. Read the local newspapers. Watch local TV in your hotel bedroom. Go to cafés and bookshops which may prove to be a good source of information.

As a business woman you may well have to travel alone. Lone women are prey not just to sexual advances, but to advances from the lonely and over-friendly. Do not spend time with people who don't appeal to you. Stay away from them.

Choosing accommodation

Ask advice from other business women who have travelled alone. Check the hotel's attitude to women guests. Some hotels allocate entire floors to women only, for example, or ensure that women are roomed near the lift to avoid long walks down poorly lit corridors.

If you stay in a hotel chain, you forfeit character and charm for the guarantee that the hotel will conform to a particular standard wherever it might be in the world: decent sized, soundproofed rooms, room service, phone, fax and telex facilities, drycleaning service and the advantage of a wide range of eating options, from grill, coffee house and dining room to decent room service.

But smaller hotels are friendlier, and often provide a more memorable stay.

The hotel's location is important. It should obviously be in a convenient part of town for your appointments, particularly if you have heavy samples or a weighty briefcase. It should also not be in the middle of the red-light district or in a part of town notorious for bag snatchings.

When booking-in ask for a room on the quiet side of the hotel and ask to see it. Look carefully at your room when you are taken to it. This is the time to ask for a different room if you are not satisfied. Before you unpack, check the plumbing, air conditioning, TV/radio, fridge, minibar, security chain. If these items are not in order, don't be shy about complaining. Ask if there is a safe in your room or safe deposit facilities on the premises. Leave extra cash/cheque cards here for safe keeping.

Some noise is inevitable in a building where many people are staying but you do not have to put up with parties or loud TV. Telephone reception, complain, and ask them to do something about it.

So what should the business woman traveller look for in a hotel? Consider the following services and decide which are essential for an enjoyable and productive trip.

*Located in a safe area.
*Safe room with security chain and well lit public areas.
*Laundry.
*Hair-dryer, TV, radio, fridge/minibar, travelling iron.
*Telex, fax, typing services.
*Pool/gym/sauna for relaxation.
*Parking.
*Safe deposit box.
*Women's facilities. Trouser presses are found in most hotel rooms these days, but it can be difficult to get an iron.

If you will be travelling around on your visit and need to find hotels as you go make sure you arrive at your next venue at a suitable time of day. Have a good guidebook which lists hotels in terms of category and facilities. If you find yourself in difficulties, contact the local tourist office.

Think bags!

Presentation kits, samples, artwork or documents can mount up to an enormous weight of baggage. Do not travel alone with more than you can carry by yourself. It makes you vulnerable if you are unable to manage without help. Remember you can send goods on in advance.

Take a hard look at what you intend bringing. Is it all necessary? Could it be condensed? Brought in a different form - could you bring photographs, for example, instead of the samples themselves? Could you send materials ahead by courier or post?

If you are going on a buying trip consider the possibility of sending materials home by post. Take tried and tested clothes, nothing that makes you feel uncomfortable. It is important to have enough clothes and accessories, although if you are meeting different people each day you can wear the same outfit. Most hotels have a good overnight or 24-hour laundry service.

Unpack as soon as you arrive at your destination. Take crease-free clothing as much as possible. Again, it may be possible to get an iron from the laundry section of the hotel, but don't bank on it.

See the regional information section for tips on clothing to wear in different countries.

Remember that luggage goes missing regularly from airlines and hotels. Your hand luggage should contain cosmetics, nightwear, underwear, toothbrush, hairbrush, medicines, anything valuable or irreplaceable, and anything that you would find indispensable for your visit. You should also wear a money belt which can contain cash and travellers' cheques.

Make sure your insurance policy has sufficient limits to cover your needs, including expensive samples and the smarter clothes needed for meetings and functions. Don't forget to take a photocopy of your passport and all essential documents such as medical prescriptions and leave them at home in the hands of someone who can forward them to you in an emergency. Also leave a copy in your hotel.

Covering your absence at work

Before you go divide your work into three categories:
> must be done
> should be done
> can be left optional.

Delegate anything that can be delegated, then work through your three categories. Brief whoever must be briefed about your absence and what might happen while you're away. Make a list of anything you anticipate happening together with action you want taken. Indicate others in the company who might help. Note the things about which you want to be personally consulted. Indicate 'extra' jobs, eg filing, which you would like to see done while you are away.

Leave full details of your itinerary plus a copy of your passport, important documents, contact address and telephone/telex/fax numbers with a member of staff.

If possible try to have your incoming mail sorted into three piles: 1. needing urgent attention. 2. non-urgent. 3. to be read at leisure.

Covering your absence at home

If it is to be a long trip arrange for bills to be paid in your absence. Make your home safe, eg central heating pipes cannot sustain long periods at sub-zero temperatures. If possible, get somebody to house-sit for you in your absence. They will be security for the house and can also look after your plants and keep your home intact while you are away. Stop deliveries of milk and newspapers. Make sure you have a home insurance policy to cover burglary.

Check diary for events such as weddings or birthdays which may occur while you are away. Prepare a card or present and entrust it to somebody else to post.

Leave details of where you can be contacted.

See page 40 for advice on coping with leaving children behind.

Avoiding travel stress.

*As a business traveller you need to keep a clear head. Start your preparations well in advance.

*Don't leave everything to the last minute and don't work late on your last day in the office. Have a relaxing evening and a good night's sleep.

*Don't travel at the end of a working day.

*Don't underestimate travel time to the airport and between airport terminals in larger international airports.

*Don't plan ahead to have a meeting immediately after you arrive. Give yourself a reasonable break, long enough to book in, unpack, wash, change and relax.

*Make sure you have the direct dial code for home when you leave. It may be difficult to get it abroad, especially if there is a language barrier.

*Make a list of the contacts you need to make abroad. Check you have correct telephone numbers/addresses before you go.

*Make sure your credit limit on credit cards is valid and will cover you in case you need to spend in an emergency.

Retired older travellers

Travel is the ideal retirement activity, providing new goals to replace the career aims, or family organisation, that once occupied so much time. Travel also helps to keep one physically fit and mentally alert.

Some car hire firms and adventure holiday companies will not accept holiday makers or travellers over a certain age. On the other hand some countries have special travel facilities for the :retired. And for many citizens travel becomes cheaper once they reach retirement age and can travel outside peak times.

The biggest advantage of travelling when you are older is that you have time. You are no longer required to rush back to a family or a job. And time is a precious resource which can make travel not only cheaper, but a lot more enjoyable and rewarding.

However, it is important to realise why you are travelling and to accept that you are no longer as young as you were. Whatever your reasons for getting on the move, your aim is to enjoy yourself, so establish what you want from a trip, then analyse honestly your ability to cope.

Travel can be tiring, so don't push yourself hard. Instead of staying one night and moving on, why not stay two or three? See if you can break up your long-haul flight without incurring extra expense. These are the wonderful options that only somebody with lots of time can avail of.

If you suffer from a medical condition make sure your destination can meet your needs. And ensure that your insurance policy will cover all eventualities. Check the policy for age-related exclusions or excesses. Airports and customs can be very confusing for older women unaccustomed to travel. Ask your travel agent/relatives to describe the process and the stages of going from the check-in desk to the plane itself.

Remember that the older you are the longer it takes to recover from illness or injury. Make a realistic assessment of your fitness and don't take unnecessary risks.

Be particularly careful of altitude. If you are going to one of the world's high spots, get a medical check-up before departure and make sure you get plenty of rest on arrival. Even if you feel well, resist the temptation to run out and see the sights.

The same applies to extremes of heat or cold. Recognise that it is less easy to heal when you are older. And be careful. Get yourself fit before you go and be good to yourself on your way.

Special services for the retired

Older people are an attractive prospect for the travel industry, as their flexibility allows the industry to fill vacant airline seats and hotel beds during low season. As a result many travel companies now provide specialist options for older travellers.

Many operators offer packaged long stays in tourist destinations at amazingly cheap prices. These usually take advantage of the quiet winter season, and can be a very attractive alternative to spending the winter at home.

There is also an increasing number of holidays which combine smooth travel arrangements and efficient back-up service with independence once you reach your destination.

Disabled women travelling

You are a disabled woman, and you want to travel the world. You can. There is no doubt that having a disability of any sort will increase the challenge of travel. Even in areas where your needs have been taken into account, disabilities like blindness, deafness, diabetes, heart conditions, needing a wheelchair and so on will present specific problems. But the good news is that it can be done and that most tour operators are now giving more thought to the needs of disabled holiday-makers than they did in the past.

Planning your trip

The countries which are best equipped to accommodate travellers with a disability are the US and Canada, Northern Europe, Hong Kong, Singapore, Australia and New Zealand.

Allow plenty of time to do research before you travel. The tourist office of the country you wish to visit will be able to send brochures, maps and information leaflets. They may also have specialist information about facilities for disabled travellers as well as weather details, health requirements, and details about places of interest. Also consult travel agencies, brochures, guide and travel books.

If you need to take prescribed medicines abroad with you it is your responsibility to know what you can take in and out of the country. Carry a letter from your doctor giving details of drugs prescribed, remembering that some medicines available over the counter in your own country may be subject to control elsewhere.

Take your normal medical needs with you, and perhaps a

little more to cover you for emergencies. Go on the premise that you will not be able to get what you need abroad.

If you are travelling abroad for the first time you may feel happier carrying a card giving your name, nationality, language, disability, type and dosage of medicine, any allergies and the telephone number for contact in an emergency.

If you use a wheelchair remember to bring a repair kit, with spare tube, puncture repair kit, pump, tyre levers, screw driver, spanner, pliers and oilcan.

Check-list for disabled women travelling

Make a listing of the facilities you are likely to need. This should be useful for your travel agent or anybody who will be involved in the organisation of your trip. It could include:

*Accessible transport to and from your home to your holiday destination/port/airport.
*Assistance at the airport/boarding the plane.
*Accessible accommodation: is there ramped or level access? are there lifts? what about parking? width of doors? accessible toilet and bathing facilities? height of bed?
*Accessible amenities, transport? shops? entertainments? beaches?
*A hospital that accepts transient patients for kidney dialysis.
*Restaurants or hotels with braille menus and safety instructions. An airline that permits guide dogs in the cabin.
*Hotels with safety equipment geared towards the deaf.

Make sure that any courier arranging transfers is aware of your disability and your special needs.

Should you choose group or independent travel? That's really up to you. Travelling with a group has the advantage of having transport, accommodation and outings organised around your disability. Volunteer helpers are usually provided, and a lot of the worry and hassle of travelling is minimised. Group travel is particularly attractive to the severely disabled.

Travel options for:disabled travellers

Air

At the time of booking your flight inform your travel agent or airline of the kind of assistance you will need. The day before departure check directly with the airline reservation desk, the special facilities desk or the customer services department to make sure your special needs have been fully understood. Also make sure that the airport staff at your destination, as well as at any stop-over where you will have to disembark, are fully aware of your needs.

Airlines differ in their policy on carrying electric wheelchairs. Criteria usually are that the wheelchair folds and that any batteries are emptied of acid or enclosed in an acid-proof container. Mobility in an aircraft is not generally possible in a wheelchair, although some airlines stow a transit wheelchair which is narrow enough to go down the aisle.

Rail

Rail facilities for the disabled vary enormously from country to country, with the best facilities to be found in the US and parts of Europe. Travelling across any big city by underground rail is not usually a viable option for wheelchair users.

Road

If you can drive either a conventional or a hand-controlled car you can easily go independently. Hiring a car with hand-controls is also an option if you are flying to your destination. Car hire companies usually require at least three weeks advance notice to organise a hand-controlled car. You should specify the size and style of car you require and whether you want the controls mounted to the right or left of the steering wheel.

Also check what sort of insurance and car documentation is required to rent and drive the car. Get confirmation of rental conditions and hand-control requirements in writing before you travel.

In the US it is also possible to rent vans and travel homes fully equipped for disabled drivers or passengers.

Sea

Taking your own car abroad on a car ferry is also a good option. Most ferry operators ask for prior notification and details of the make, colour and registration number of your car, the number of your seat or cabin reservation, whether you are a wheelchair user, and details of assistance you may need, including lifting up stairs if lifts are not available.

Information about accessible facilities offered by the different shipping companies may be hard to come by. Travel on personal recommendation if you can. Usually more modern ships run by larger operators tend to be better equipped than those of smaller companies. Put details of your needs in writing and retain a copy of the letter.

Most ferry companies ask disabled drivers to arrive an hour or more ahead of sailing time so that they can be placed at the front of the vehicle queue. Tell the ferry staff immediately on arrival at the dock that you are a disabled traveller needing assistance. They will radio ahead to the ferry loading officer to make sure someone is standing by to direct you.

Taking a cruise is another option which has recently opened up for wheelchair users. A cruise is also a restful option for those suffering from heart conditions. A good travel agent should be able to tell you which shipping companies operate cruise holidays with suitable facilities for disabled passengers, or you can contact the cruise shipping companies direct. Most ships are equipped with the same basic features, but the design can vary considerably. Ask for a deck plan.

Find out if any cabins have been designed for or allocated to disabled passengers; whether passenger lifts on the ship are large enough to take a wheelchair; whether they provide access to all decks; whether the entrances to restaurants and other public areas are barrier free and whether motorised wheelchairs are permitted on board.

Need special aids?

If you will need aids on your trip and cannot take your own, contact the branch of the Red Cross Society nearest your destination, who will advise you on the hiring of aids.

You may prefer to invest in some essential equipment that is suitable for both home and holiday use. Items should be chosen for ease of packing and transportation. Examples include a collapsible commode for times when the toilet is inaccessible, a set of folding rails to place around a low toilet, a set of bed blocks to raise a low bed. Wheelchair users who intend doing a lot of sightseeing should bring a sliding board and an extra pair of pushing gloves.

Many aids are not included in baggage weight allowances (check with the airline before departure, though) and medical equipment may be taken on airplanes free of charge.

A female urinal is a valuable incontinence aid for travelling. Some women find them impossible to use, others use them with great success. There are several types available. Go to your local aids centre and check out which might be of most use to you. Whatever aid you use, it is best to practise before you go until you become completely familiar with it.

Travelling when pregnant

If you can choose your time for travelling go between weeks fourteen and thirty. At this stage you are less likely to be troubled by morning sickness, you won't be too large and you shouldn't be too tired. The first three months of a pregnancy are the most risky and the final three months also pose health risks. Find a sympathetic doctor or midwife, one who has travelled or who deals with travellers, and ask advice.

When planning your trip bear in mind that some airlines have regulations regarding the last few weeks of pregnancy, when they refuse to carry you, and that some insurance companies will add caveats to your policy relating to your condition.

When you book your ticket inform your carrier that you are pregnant. If you are beyond twenty-eight weeks, they may require a doctor's certificate of fitness to travel. Many airlines will not accept you for international journeys after thirty-five weeks. Read about the countries you are going to and choose your route carefully.

Think ahead when packing. From twenty weeks - probably earlier on a first pregnancy - you will increase in size and many women find they cannot tolerate tight clothes from a very early stage. Bring only loose-fitting clothes and comfortable, roomy shoes. Pack a good book on pregnancy.

Vaccines and medication

If you are planning to become pregnant try to get all your immunisations before you conceive. Whether you are able to do this will depend on your destination, as the periods of immunity afforded by some vaccinations are limited.

Drugs which have a harmful effect on a developing foetus do most damage if taken during the first fourteen weeks while the main organs are forming. Avoid all unnecessary drugs during this time. Otherwise it is a matter of balancing the risk of taking the drug against the risk of contracting the disease.

There are two types of vaccine: dead vaccine, such as that for cholera, and live vaccine, such as the yellow fever vaccine. Live vaccines should be avoided in pregnancy. If you are going to a country where a certificate of vaccination is required and your doctor advises against being vaccinated, get a letter from your health authority saying inoculation is contra-indicated. This is generally accepted.

Vaccines which normally cause fever, such as diphtheria vaccine, should also be avoided.

Malaria is highly dangerous to your unborn child as well as to yourself and when you are pregnant malarial attacks tend to be particularly severe. There is a high risk of death or of losing the baby, so travel to malarial areas during pregnancy should be avoided.

If that is not possible chloroquin and proguanil are the safe anti-malarial drugs for use in pregnancy, but if you are using either, take folic acid tablets with you, as the drugs will deplete your natural supplies. Remember too that there is a growing increase in the spread of drug-resistant malaria through most of the countries where malaria occurs.

Tetanus injections are safe in pregnancy and the immunity is passed on to the baby. Some antibiotics, however, may not be safe. Tetracycline, an antibiotic prescribed for cholera, can lead to discolouration of a baby's teeth. Streptomycin can damage the nerves in a baby's ear. Always inform your doctor about your pregnancy and discuss beforehand the implications of taking a drug.

Ask your doctor about taking an injection of immunoglobulin before you go, as protection against Hepatitis A.

Two other commonly used drugs which should be avoided in pregnancy, particularly in the final weeks, are phenolphthalein, a laxative, and aspirin.

Find out as much as you can about local medical care: the names and addresses of doctors, hospitals and facilities for intensive care in case anything should go wrong.

On the move

If you plan to be away from home for a long time, try to continue your ante-natal care by visiting local hospitals at monthly intervals to get your weight, urine and blood pressure checked.

Don't ignore bleeding in pregnancy. Rest immediately and call a doctor. Rest for at least three days after any discharge has ceased. If you have bleeding accompanied by pain in the lower abdomen, or a heavy bleed, you should seek medical help urgently. It might be an incomplete miscarriage and remaining membranes could cause life threatening blood loss and infection.

Severe swelling is another danger sign, particularly if the face and eyes puff up. Go to a hospital and get them to test for protein in your urine to ensure you are not suffering from eclamptic toxaemia. If you are you will need bed rest. If this condition is not checked it leads to headaches, blurred or flashing vision and, in the final stages, fits. It is potentially fatal for mother and foetus.

Don't push yourself too hard on your travels. Listen to your body and don't expect to react as you would when not pregnant. Arrange your schedule to suit how you feel. If you need a nap in the afternoon, for example, get to your hotel by lunchtime.

On long journeys you may find your feet and ankles swell, and this will be aggravated if the weather is hot. If possible, get up regularly and walk around. If not, stretch out your legs regularly and do little foot exercises, rotating your feet from side to side and flexing and pointing them.

Yeast and fungal infections are a problem for many pregnant women and travelling in a hot climate exacerbates the problem. Wear loose cotton dresses rather than trousers and wear cotton underwear.

Use your common sense; you will have plenty of time for travel after the baby is born so don't do anything you might regret. Of course, most types of travel are perfectly safe during pregnancy; it's a case of not taking on a climb up the Materhorn in your thirty-eighth week.

Food and drink

Learn about nutrition and ensure that you are meeting your nutritional needs. Be aware not just of the type of food you should be eating, but also the way it is cooked; over-cooking, for example, rids food of nutrients. In some countries dairy foods are not always available and you will need to eat other foods which are rich in calcium.

A multi-vitamin and iron, calcium or folic acid supplement

might be a good insurance policy. Check with your doctor.

It is important to drink plenty and often, in order to avoid constipation. For the same reason you should eat plenty of fresh fruit and vegetables and fibre-rich cereals. Avoid alcohol except in small quantities.

Diarrhoea is more difficult to cope with during pregnancy, when you are trying to avoid unnecessary medication. Rest and a bland diet will usually clear it but if it persists you might want to try a kaolin preparation (get it before you go).

Despite all the warnings above remember that pregnancy is not an illness and that most pregnant women travel without any untoward effects. But forward planning and care will help you relax and enjoy that much-needed break before the arrival of your child.

Travelling with children

Children change everything in life and travel is no exception. If you bring them with you, every decision from where you go to what you bring will be affected. If you leave them behind that carries its own set of problems.

Let's look at bringing them first. The initial thing you have to accept is that you will not enjoy your trip unless they are happy. But that doesn't mean you have to restrict yourself to holiday camps or beach breaks until they are eighteen. Children are good copers; what they are brought up with is what they take to be the norm. They quickly develop a hardy tolerance, even a liking, for unconventional holidays if that is what they are used to.

If you are determined to make things work, there is no kind of holiday on which you cannot take children.

Why bring them?

With all the extra worries and hassle involved in bringing the children away you might ask yourself, why bother? Would you not be better off if they stayed at home with fond minders while you enjoyed a break from everything, including them?

Well, perhaps. But, aside from the difficulties of finding fond minders, many parents discover that, despite the hassle, bringing children along makes their trip more rewarding. People treat you differently. When you meet local people with children in tow you immediately have something in common. Other children come up to look or to play and it's easy to strike up communication with parents even if there is a language barrier.

Most of the things you want to do when travelling will have some appeal for children, if not quite the appeal you expect. They can help you see a country in a new way. It's also good for them. They should learn a lot, and become more resourceful and self-reliant.

It's good for the family, whatever form your family unit happens to take. Travelling together gives an added dimension to family life; for nuclear families, single parents or happy couples who also have children along, travel can provide a sense of togetherness. You have time to be together without having to rush to school or work, and domestic responsibilities are left behind. Even a young child who seems to have no conscious memory of a holiday draws the benefit of this.

The challenge is to balance what you want from the trip with what you hope your children will get from it. This means compromise, which really means *you* compromising. After all, the trip is your idea. The kids would be happy at home. If you want them to be positive about travel you really have to plan your holiday around their schedules.

Endless days of sightseeing, therefore, are out. But a morning viewing the sights followed by an afternoon on the beach can be enjoyed by everyone. Above all, it is important to be flexible and to remember that your objective is everybody's enjoyment.

Before You Go

Travelling with children means that advance organisation is not an optional extra; it is vital to the success of your trip. If it is your first trip since your children were born, they must be included on your passport or have passports of their own. Allow at least six weeks for this. If you are travelling with a partner it is worth putting the children on both passports in case for some reason one of you might have to return home early.

Get as much literature on the destination as you can - tourist brochures, books from the library, anything with pictures. Find books of legends or children's stories from the region. For older children there are some excellent books on many countries. Older children may also enjoy looking at maps, working out a route, or learning simple words in the language.

Include the children in planning the trip; ask them what they want to bring. Take them to eat in restaurants, for walks, and on short day trips for some weeks before you go, to prepare them for travelling. If you have a baby get her or him used to cold bottles and jars of babyfood before you go, and if you are bringing your own travel cot, let the baby sleep in it for a couple of weeks before you leave.

Check with your doctor about health precautions for your destination well in advance of departure. (See 'Children's Health', page 39). Ensure that you take out holiday insurance well ahead of departure date.

Questions to ask about your destination

*Where does your prospective accommodation stand in relation to other facilities? How will you get around? If you have young children and are taking a pushchair, are roads steep/level, in good/bad condition?
*Are there facilities, sports or entertainment on offer for different age groups? What about baby equipment like highchairs, safe cots? If you intend to let your children swim there, are waters clean and safe? Is the beach sandy or stony? Is there a play area? A children's club?
*Are there any hazards such as strong tides, jellyfish or sharks? What about busy roads which might be noisy at night? All-night discos? On what floor is your bedroom? Is there a lift? Are balconies safe for young children?
*How well-stocked are local shops? Do they sell disposable

nappies? Baby food? How much do they cost? Is there a laundry or nearby launderette? A baby-sitting service?
*If you are hiring a car what about child safety seats?
*Who do you contact if you have problems with travel arrangements?

The journey

The journey is an important part of your holiday and you must be well-equipped and prepared. Due to increased security, airlines now insist that outsized bags are unacceptable. Check this out and pick one standard sized bag for the journey itself so that you do not have to rummage through the rest of your luggage. Choose bags that are easy to carry, either rucksack style or bags with wide straps that don't cut into you.

Unless you are travelling alone with children distribute the load as much as possible: have one bag for cleaning equipment, one for food and goodies. Give the children a bag of their own, allow them to carry their own toys and games, and a spare T-shirt or jumper. This gives them some independence and responsibility.

Adults too should take a spare T-shirt or jumper to change into on arrival, particularly if travelling with a young baby.

Games and toys are not a luxury on a journey. But do dissuade the children from taking anything cumbersome or heavy. Also avoid toys that make irritating noises, such as whistles, loud rattles, or bleeping computers. Anything messy like markers or plasticine should also be avoided. Most airlines provide packs of colour books and crayons for children, but do check this out.

Best toys for travel would be books, cards, magnetic games, travel sized versions of standard games like draughts, chess or scrabble, puzzle/colouring books with crayons or coloured pencils. For babies bring teethers, mobiles, building beakers and board books. Bring personal stereos, a must for teenagers who can also use the ear plugs for audio equipment on planes. Don't forget to bring some biscuits or things to chew on the journey.

Small babies rarely seem to suffer from travel sickness, but once they are toddlers and beyond, watch out! Sickness is caused, in part, by the discrepancy between the information received by the brain from muscles and joints, which say we are at rest, and that received from eyes and ears, which say we are moving. Tell the children to avoid looking out the side windows of a moving car, for example. Remove the head rests to increase forward visibility, and sit small children on raised seats. On a coach sit over the coach's centre of gravity (in the centre and not over the wheels) to reduce rolling. Discourage the children from reading.

A light meal an hour or so before setting off is good, but avoid greasy food before and during the journey. Dry, plain biscuits can help. Avoid the anxiety induced by a last-minute rush. On the journey regular stops for fresh air help. Carry headache tablets in your bag, just in case.

If all else fails, drugs can be effective. They dampen the sensitivity of the balance mechanism in the ear. Despite precautions, be prepared. Have plenty of tissues or wet wipes to hand, as well as a receptacle for vomit.

Taking children on airplanes

Young children are hard going for a long flight. Changing nappies in a airplane toilet and trying to eat airline food with a baby on your lap is not fun. Small babies will sleep a lot, but from age one to three it is very difficult to keep them amused. From three years of age onwards it gets easier. Simple games amuse them for a while and they can eat unaided. Do spare a thought for sister travellers who may not have the energy to cope with your children. Keep an eye on your children and don't let them annoy other travellers.

If you are travelling with an infant, inform the airline when booking. Some airlines have a 'skycot' for young babies, useful also for very young children (average dimensions 60cm x 30cm x 20cm). Pillows and blankets are also available, although usually only handed out on a night flight. If your child has a favourite rug or blanket, bring it with you if possible.

Check in early to give yourself the best chance of choosing a seat. Insist on choosing the seat you want; look at the plan of the plane and pick a suitable seat which gives you plenty of room. Most airports have a parents' room where you can change and feed a baby. Some are excellent, some so useless it's hardly worth having them at all. Few airports cater for older children. Travelling parents should put pressure on airports to provide facilities. It can be done. Singapore's Changi Airport has a wonderful play area for children of all ages.

Check the facilities in the departure lounge before you go in. Phone the airport and get information on the facilities from the public information officer. Some may not even have a snack bar and once you are through you will be unable to return to the main body of the airport.

Make sure younger children visit the toilet near boarding time, as you may get caught in a queue later. Most airlines allow families to board first. Most also allow you to wheel your pushchair to the departure gate, where it will be taken and stored in the hold until you arrive.

Children under two years are not usually allocated a seat, but are expected to sit on your lap. Neither do they have a baggage allowance, despite the fact that their equipment is often the heaviest; however, you may be allowed a carry-cot and pushchair in addition to your own baggage allowance.

A seat by the bulkhead (the partition separating the aircraft compartments) is probably the best choice, if you can get one. They give more space and your child can kick and wriggle without disturbing a passenger in front. The extra space at your feet may even be made up into a bed on the floor. A window seat gives most privacy for breastfeeding.

Airlines usually make an effort to give families the benefit of spare seats if a flight is less than full. With a couple of seats to stretch over small children can sleep quite comfortably.

Young children are very sensitive to pressure changes at take-off and landing. Sucking and swallowing help release the pressure in the ears, so it is a good idea to give your baby the breast or bottle at these times. Drinks with a straw or sweets to suck are best for older children.

Some airlines provide food for children, but bring your own food too. Jars are best. If the crew heat the jar or bottle for you make sure it is cool before you attempt to feed a baby on your lap. Turbulence could make you spill the food or drink, causing a nasty burn.

If your baby is breast fed, bring some juice in a bottle to give yourself a break as everybody gets dehydrated on planes, including baby. And watch your own fluid intake. Use the water fountains regularly or drink mineral water.

As for any other journey have a good supply of toys and books and bring them out one at a time. Provide yourself with a head-rest or plane pillow for comfort.

If you are crossing time zones set your clock to your new time when you get on the plane, and think in terms of your new time from that moment on. You will all probably feel like falling into bed when you reach your destination. Don't. Take it easy, relax and wind down, but try to make bedtime appropriate. Let the children nap, but keep the going-to-bed rituals (nightclothes, bath, story) until new bedtime. Invest in a world clock; these are quite cheap and can be useful when you are away and want to call home.

If you do make a long trip, spend the following few days in a comfortable place where the children can run around and enjoy themselves and get the trip out of their system. Never plan too many long journeys and where they are unavoidable, space them as much as possible.

Taking children by boat, train or coach

If you are travelling overnight by boat or train, you should book sleeping accommodation. Night travel with a sleeping compartment is a good idea when you have children. Yes, it is more expensive, but if you balance the cost against the cost of finding alternative accommodation, and the cost of arriving at your destination exhausted, with fractious children, it can be well worth it.

If you take a car-ferry remember to carry a bag with a change of clothes for the children, as well as a towel, tissues or a roll of toilet paper which can serve as both, toys and books and anything else vital to the enjoyment of the journey. Once the ferry has left port you may not be allowed to the car for such items.

Boats and trains are the most suitable forms of transport for parents with children, if only because they give the children enough space to move around. Have alternative sources of food with you, though. Cafeterias and buffet cars may be closed, full of queues or serving inedible food. Bring an adequate supply of drinks too.

The only reason for choosing coach travel is the price. It can be very much cheaper than air or rail travel, but it is comparatively slow and often uncomfortable, which means that, if you can afford to, you should avoid coaches when you have children in tow.

While on holiday

Structure your day around the children's needs where possible. Don't fill all of every day with things to do. Children need some unstructured playtime each day. Keep them informed of plans. Involve them. Help them choose postcards. Let them have local money.

Other people's interest in your children is usually welcomed, and is a great way to strike up conversation. Many Asians consider children to be communal property and will have no reservations about kissing and cuddling your children and passing them around from person to person. Your children may start to protest if unused to this.

You need to decide in advance how to balance your wish to be courteous with your concern for your children's feelings. If you feel that you can't persuade the children to enjoy or accept the unaccustomed attention, you may have to refuse to allow them to be handled, though this may be interpreted as rudeness.

Also in Asian countries babies are not allowed to cry for long. You might know that your baby always cries for a little while before dropping off to sleep but you will be expected to do something about it, and if you don't you'll get disapproving looks, and quite possibly a lecture and demonstration of how things ought to be done.

Children's reactions to the strangeness of everything may vary from sheer delight to a defensive 'this is boring' attitude. Don't try to force them. Give them lots of extra coddling, keep them with you and explain everything that is happening to them. They should soon regain their sense of security.

Meeting children's needs

*Cloth nappies are non-existent in many parts of the world. Bring your own disposables if possible. Although they are bulky, they are light, so they don't affect your luggage allowance. Multiply the number of nappies you use per day by the number of days away and add a few extra for emergencies. Use the bag you bring them in to bring home any purchases you make. Alternatively, bring about a half-dozen cloth nappies and wash as you go!

*Do try to arrive at a destination early in the afternoon to give your children time to settle in. Let them have time to explore the room, play in it, learn where the bathroom is. With older children set out toys and books. Let them play house. With younger babies have something familiar within view.

*In international style hotels you won't have much trouble getting cots. Children from eighteen months can go into a bed, but for very small babies, it is probably best to use a portable carry-cot. You can, of course, bring a child to bed with you at night, but they are likely to want to retire a lot earlier than you. And what about nap time? Bring a waterproof sheet for small children to protect hotel bedding.

*If you encounter an unfamiliar toilet, a squat-down variety when your child is used to the sit-up type, for example, try to make it an adventure. Show them how to use it and treat it as an interesting novelty. And hold them for the first few times they use it.

*The best way to feed a baby on the move is by breast. Bottles are successful only if you are going direct to a destination and staying put. Otherwise it is almost impossible to keep bottles sterile and formula fresh and at the right temperature. In many countries even finding clean water can be a problem. So, for the travelling woman, breast certainly is best. All the advice that applies to nursing mothers must be adhered to while travelling. Slow down, take things easy, drink lots of fluid and make sure you eat properly.

*Hygiene rules apply even more strictly for children's less hardened digestive systems. In parts of the world where water is not safe to drink, small cartons of fruit drinks or bottled water is good for children. Always carry a water bottle with boiled, purified water. Milk and other dairy foods are not widely available in some countries.

*Children are notoriously unadventurous in their eating habits and you may have to pass up some wonderful local eating opportunities while you search for a place that can produce a mundane hamburger or hot dog. Give in occasionally at least.

*Don't forget room service, particularly for those nights when you are just too tired to face a restaurant. Even a cheap hotel will be able to send somebody out for you, or one of the adults can go to a local take-away or fast food place. Do anything that will fill their stomachs without subjecting you to unnecessary stress and fights.

*Probably the best way to carry an infant is in a front sling. The baby is warm and comfortable and soothed by the beat of your heart. It's as close as they can get to being back in the womb. However, babies soon outgrow these slings and by three or four months can be too heavy. Once they can sit up by themselves a backpack is a good alternative. There is now a backpack-cum-stroller on the market which is a good idea if you travel a lot, because you can sit the baby in for meals or for a nap.

Children's health

The list of potential dangers can seem frightening but with basic precautions, adequate information and a little luck, most travellers (and their travelling children) experience little more than a stomach upset.

When you are abroad it is much more important to watch what children put in their mouths than it is at home. Let them know the dangers of accepting sweets and drinks from others in regions where lack of hygiene may cause your system problems. Do not allow them to pick up and eat anything that has been dropped. Ice lollies and ice cream may not be safe; they may be made with unclean water. Avoid allowing the children access to these as treats.

Impress upon your children the importance of not petting animals. Dogs, cats and other mammals such as monkeys are all capable of transmitting rabies. Stray animals should be avoided.

Children should always wear shoes or sandals outdoors, to prevent burning on hot sand or soil, or in the tropics to avoid the nasty infections such as hookworm or jiggers that can be caught by walking barefoot.

Dehydration can be a real danger for children. Always carry a water bottle with you and give frequent drinks. Don't rely on them feeling thirsty to indicate they should drink; check their urine. Not needing to urinate or very dark coloured urine are danger signs. Another simple test is to pinch a small portion of skin on the soft part of the forearm. Hold it for a second, then let it go. If the skin goes back slowly, dehydration may be the cause.

Children often have high fevers for little apparent reason and recover remarkably quickly. Carry a thermometer and fever strips. The general rule for coping with a high temperature is to get the child cool as quickly as possible. Remove all clothes, sponge the child with cool water, place close to a fan and get the child to drink cool drinks. Administer paracetomol. If the temperature continues high, place the child in a cold bath (ignore protests). Get medical help if the temperature is abnormally high or if it persists.

Make sure children are up to date with their routine immunisations programmes before you travel. If you are travelling to a malarial area, children too will need an anti-malaria drug, one week before they travel and for a month afterwards. They will take a reduced dose. Your doctor will advise you.

Children over six months can be safely inoculated against yellow fever, cholera, typhoid and can receive protection against Hepatitis A. Again, check with your doctor.

Suggested medicines for small children

*Junior paracetomol or other pain-killing syrup.
*A rehydration mixture for the treatment of severe diarrhoea. Available in sachets, the most convenient form for travel. Mix with bottled water.
*Calamine lotion, to ease irritation from sunburn, bites or stings.
*Cream for nappy rash.
*Teething gel.
*Thermometer/fever strips (mercury thermometers are prohibited by airlines)
*Travel sickness pills appropriate for the age of your children.

Leaving the children at home

If you decide not to take the children along the first essential is to provide proper care for them while you are away. Don't compromise on this even if it involves trouble and expense. If you are not happy with the child care arrangements you won't have an enjoyable holiday or productive business trip.

Make sure your minder is fully briefed. What does she do in the event of an emergency? Is there somebody nearby, a partner, relative or friend who can help her out if things go wrong? Of course you must leave your forwarding address with her and details of your itinerary if you are on the move.

Explain to the children in age-appropriate terms where you are going and why, and how long you will be away. If they are old enough to be interested involve them in your trip. Show them brochures or books about your destination. Be matter-of-fact in your explanations; if they sense that you feel guilty or upset about the impending trip they will feel confused and may become upset themselves.

Try to ensure that you don't go at important times like birthdays. And, of course, remember that the best way to resign children to the prospect of your absence is to mention the magic word: presents. Emphasise what you will do together when you return and what you will be bringing back and you will probably find them looking forward to your trip.

For your own part, don't tell yourself that you won't miss them. You will. But keep a sense of perspective. In all the days and weeks and months and years you spend together a short break from each other doesn't count.

The more often you go, and the older they get, the easier it becomes.

You want to go where?

For a short trip, the decision of where to go is relatively easy. Money and time will generally determine your destination, along with other factors like ease of access, and perhaps fashion. Those planning a longer trip will have more choice.

Ask yourself the following questions:

*Are there any countries with a culture that interests me and makes me feel I'd like to find out more?
*Do I have friends or relatives living in a part of the world I would like to see?
*What is the status of women in countries I might visit? How might this affect me?
*Do I want the challenge of experiencing something very different from the country I live in or do I want a relatively familiar environment? Would a country sharing my language be more attractive to me?
*Do I have particular goals in mind, or things I want to learn about myself? What places might best achieve those aims for me?
*What about comfort? Do I feel I require a certain standard to enjoy myself? How might that limit my choice?
*Do I know any other women who have travelled widely? What can I learn from their experiences?

Getting in the know

The more you know about the country you are to visit, the more you will get from your trip. So, research your destination. Guide books are one good source of information, as are travel books. Write to the relevant tourist office; they are likely to send you lots of colourful literature outlining the attractions of the country or area of interest to you. Talk to other women who have travelled in the region.

The National Geographic magazine is an excellent source of background material. Large libraries bind the magazine with an index, making it a simple matter to look up your special interest. Articles from newspapers and magazines can be cut out and taken with you. Newspapers give an update on the political situation, particularly important in areas where there might be trouble or unrest.

Bring some of your reading material on your trip. Writing which appeared only mildly interesting when read at home becomes quite riveting once you're there. When you arrive look around for material that you might not have been able to get at home. National libraries may have books in English, or another language you can read, as may university libraries.

Getting plans afoot

Think about the time available to you and how you would most like to fill it. Loosely plan your itinerary before you go, leaving plenty of time to enjoy the unexpected diversions which are half the fun of travelling. Note only the things you really want to do on the trip, and be flexible otherwise.

Don't get yourself into a flat spin wanting to do everything and see everything. Decide that you can't see everything.

When planning your trips remember that long journeys are tiring. So as well as working out what you want to do and where you want to go, build in rest time for just hanging around and doing nothing.

Some travellers make out a fairly uniform schedule: two days here, see such-and-such, then two days in the next place, and so on. But it isn't necessary always to move at the same speed. You can travel long distances fast, then slow down and spend a longer period in one place, perhaps taking day trips, perhaps just relaxing and absorbing the feel of a place which you can only do by spending time in it.

If you are taking a long trip, lasting some weeks or months, you are most in danger of giving yourself an overcrowded agenda. Travelling through a series of countries or states involves complex border procedures, inevitable delays with unpredictable public transport, not to mention culture shock. It's all part of travelling. Try to slow down, to take the pressure off, to enjoy each experience.

In Third World countries, particularly, Western obsession with the clock is entirely inappropriate. Floods in the rainy season, unpaved roads, dilapidated vehicles - these are just some of the everyday mishaps you are sure to encounter. You have no control over this, so you might as well relax, drop the deadlines, and accept that when you arrive, you'll be there. Knowing your destination's climate is essential to your planning. What you will wear, what you will eat, how you will get around will all depend on what the weather is like. In the following regional information section there is general information on climate, along with graphs of specific towns showing monthly temperature and rainfall variations.

Africa (North)

The African continent is commonly perceived as one large country. In fact it is one large land mass made up of about fifty countries. Bearing this in mind, each country will have different entrance regulations. These need to be checked out months in advance.

Climate

Varies from warm and pleasant Mediterranean climate in coastal regions to the arid heat of the deep Sahara. Coastal rains usually fall between September and May and are heavy, but not prolonged. Winter temperatures usually do not fall below freezing, except for mountainous areas which may experience some snow. In summer temperatures may climb as high as 40°C, but are usually bearable along the coast.

The Sahara, however, has an extreme climate, with

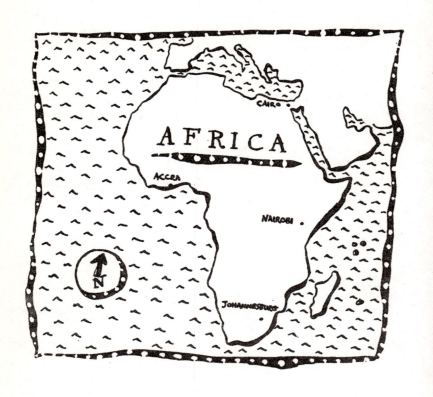

Africa

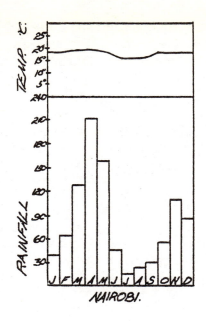

NAIROBI.

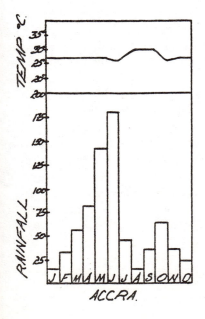

ACCRA.

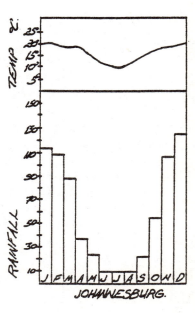

JOHANNESBURG.

temperatures varying from 3°C in winter to 50°C in summer. Rainfall is rare, nights are freezing, days are blisteringly hot, and strong winds and dust storms add to discomfort.

Northern Africa: tips

*Islam has had a major influence. As with all Muslim countries women have to be careful to observe social customs in order to avoid harassment. Wear modest, conventional western clothes which cover you well. Avoid showing bare arms (no cut-off T-shirts) or short clothes which expose the knees (no shorts, minis).

*Applications to visit Libya must be made in Arabic.

*Topless/nude sunbathing is out, except perhaps around hotel pools or beaches where guests are European and staff worldly wise. Simple rule: if everybody else is covered, don't bare.

*Eat with your right hand. Gifts should be given and received with the right hand, or both hands, but never the left alone as the left hand is considered 'unclean'.

*Hustling is a way of life. Money is often expected in exchange for an apparent 'gift'. A flower, for example, is tippable. Don't accept anything you don't want. Be firm in refusals.

*Be alert to the possibility of theft, particularly in the bazaars.

*Hitching is not recommended.

*Businesswomen should remember that Friday is the Muslim day of rest, so the work week runs from Saturday to Thursday. Social engagements begin late, with dinner often being served at 10.30 pm or later. Bring a gift and give it to your host with your right hand. It is considered impolite to eat everything on your plate.

*Any mention of time should be taken with a grain of salt. In general there is a relaxed attitude about time, but the visitor would do well to be punctual and have patience for those who are not.

Language: Arabic.

Transport: Buses and trains are cheap, but very crowded. Shared taxis operate in many areas. In desert areas trucks give lifts for a fee.

Accommodation: Finding a cheap room is easy in most areas. Only well developed tourist spots have luxury hotels. There is a good network of youth hostels throughout northern Africa.

Background reading: Mernissi, Fatima, *Doing Daily Battle: Interviews with Moroccan Women* (Cambridge University Press, 1974).
Atiya, Nayra, *Khul-Khall: Five Egyptian Women tell their Stories* (Virago, 1988).

Africa (West)

Climate

The weather in western Africa tends to be uncomfortable. Temperatures are high throughout the year and coastal areas are extremely wet and humid. There are two rainy seasons: May/June and October. It can be cloudy at any time.

In the north-west there is less rainfall with just one rainy season between June and September, but humidity is high, relieved only by the *harmattan*, a hot dry wind from the Sahara.

Western Africa: tips

*Islam in West Africa is less strictly adhered to than in the North. Local women are more evident here and visiting women are less likely to feel harassed in countries like Senegal, the Gambia or Sierre Leone in the west, than in Egypt or Morocco in the north.
*Wear long, loose, unrevealing clothes.
*Women may be approached with direct offers of sex. A firm refusal is usually enough to deter.
*Theft is common, and muggings too, particularly in cities like Dakar or on the magnificent deserted beaches which line this part of the continent.
*Hustling and begging is commonplace. Plan your strategy to

deal with this.
* While parts of Ghana and Nigeria are very populous and westernised, several other countries are not, and many goods are unavailable. Take all necessary toiletries, suncreams, books, batteries, photographic equipment and film with you.
* The people are very hospitable and quite likely to invite you to their homes though some of them can ill afford to do so. Be sensitive: do give a gift or offer a service to repay them.

Language: Various.

Transport: Varies from efficient and expensive buses and trains in Nigeria, to the cheap, irregular, crowded bush taxis of Senegal.

Accommodation: The cities and tourist resorts have a number of accommodation options but once you move out into rural areas, hotels or rest-houses are few and far between. You may be invited to stay in people's homes. Make sure you reciprocate with a gift.

Background reading: Ba, Mariama, *So Long A Letter* (Virago, 1982).

Africa (East)

Climate

The lowlands of the east have very low rainfall. Inland temperatures are kept down by high altitudes and along the coast by sea breezes. Most areas have two rainy seasons, in April or May and for a couple of months between July and November, depending on the latitude.

Eastern Africa: tips

* Sexual harassment exists, particularly in tourist areas like the coastline of Kenya, where the men seem to believe in the stereotype of western women as loose and available for sex.

*Swahili, a sort of East African Esperanto, is a made-up language, enabling people of different origins to communicate with each other. It is quite easy to read and pronounce and learning a few words from a phrase-book will be appreciated by Swahili speakers.

*Khaki is not just a fashion colour: it is worn because animals are less likely to spot you and run away if your colour blends into the background. Take a hat with a brim and sunglasses to shade your eyes from the glare of the sun.

*Don't flaunt your comparative wealth in the cities of this area. Take precautions also in resorts like Mombasa and Malindi. And don't wander city streets at night.

*Although local women may occasionally walk around topless, it is illegal to sunbathe topless in public places.

*Ask permission before you take photographs of people. A tip is usually expected.

*If giving a gift to locals, avoid flowers, unless expressing condolences.

Language: Various.

Transport: Mainly buses, although there is a small train network. Cheap.

Accommodation: Plenty of options, from government rest-houses to lodges in game-parks. This is safari country, where packaged holidays and beach resorts are available. Good camping and hostelling facilities.

Background reading: Likimani, Muthoni, *Passbook number F: Women and Mau Mau in Kenya* (MacMillan 1986).

Africa (South)

In South Africa there are strict conventions and social rules regarding race and colour, and different areas (depending on your colour) that should be avoided if you are anxious to stay away from danger. Things are changing there, but very slowly, and many women find themselves unable to visit the country at all, as to do so seems to imply tacit approval of the regime which exists there.

Climate

Climate varies from the Mediterranean climate of the Cape Province with mild winters and warm summers, to the semi-desert Kalahari, and wetter areas of the south-east. In the more northerly areas there is a rainy season from December to March when temperatures are highest, but further south there is some rain all year round.

Southern Africa: tips

Parts of southern Africa are still ravaged by war, drought and food shortages, and, of course, the notorious apartheid of South Africa. However, women travelling in this part of Africa say they feel safe, even when travelling alone.

*Hitching (usually for a fee) is widely accepted in this part of the country, and, for the most part, is safe.

*Sexual harassment of foreigners is not prevalent, but does exist. Local white men of the macho variety are sometimes to blame for forcing their unwanted attentions on visiting women.

*At dinner in some countries you may find you have to ask for food, as it is considered impolite for the host to offer food first. It is also improper to refuse food.

*As in the north, give and accept gifts and shake hands with your right hand.

Language: Various. English is widely understood in most countries. Use swear words sparingly; they are seen as very potent in intent.

Transport: Varies from country to country. Getting about independently can sometimes be difficult. Buses are slow and crowded, though cheap. Hitching is acceptable in areas of poor transport and seems to be safe.

Accommodation: Finding a place to stay can be difficult. Where hotels do exist they are geared to the tourist trade and are expensive. Council rest-houses, missions, secondary schools and lodges on game parks have all served as accommodation for independent travellers. Expect to provide your own food as these tend to be self-catering.

Background reading: Kuzwayo, Ellen, *Call Me a Woman* (the Women's Press, 1985).

Australia and New Zealand

Climate

In Australia, a rain belt follows the coast, forming a crescent shaped habitable area around the outback, largely uninhabitable. The east has rain all year round, while north and north-east have summer rains, and south and west have winter rains. Cyclones occur frequently in the north-east and north-west - Northern Queensland, Northern Territory (Darwin). New Zealand and Tasmania have a temperate, maritime climate with four distinct seasons. There are numerous rainy days, with warm summers (November to February) and cold winters (June to August). The far north of New Zealand has a sub-tropical climate; winters are mild and summers are warm and humid. In the south there are regular snow falls.

Australia and New Zealand: tips

Australia is huge - almost as big as North America - but with a population of just 16 million people, the majority of whom live on the coastline as the vast inland area is largely uninhabitable.

*Don't underestimate the size of the place. The flight between Perth and Sydney is as long as that between London and Moscow. Internal flights are expensive. If you buy your internal flights before you leave home you may be entitled to a tourist discount. Check with your agent or airline. Some

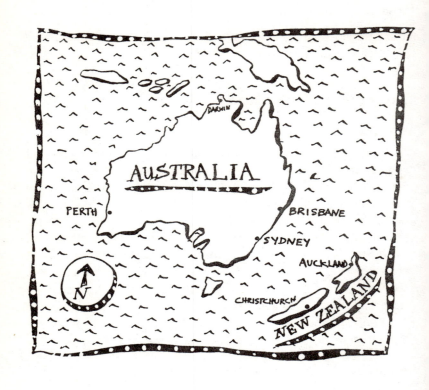

Australia/New Zealand

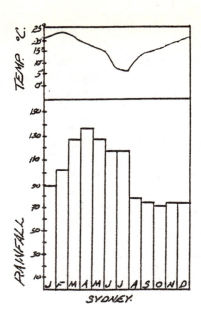

SYDNEY.

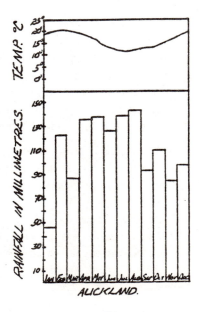

AUCKLAND.

airlines give you a voucher for two free internal flights. Check this out.
*Sexual harassment is not uncommon in Australia. In the outback - an area where hard physical work is the mainstay - macho culture prevails. Bars are segregated and a woman's place is clearly defined.
*Australians are very hospitable and very casual in their dealings. They are less formal than Americans or Europeans.
*What you wear is a matter of individual choice.
*If you are considering stopping over on your way to Australia, check the price of stopovers against the price of a round-the-world ticket which gives you a lot of flexibility and may not be more expensive.
*The sun is very, very strong in this part of the southern hemisphere, so you need a very high factor sun protection cream (15+) if you intend sunbathing or spending time outdoors. Good sunglasses are also necessary to prevent sun-related eye problems.
*New Zealand men tend to be as chauvinistic as Australians, so be careful.

Language: English.

Transport: Australia and New Zealand have sophisticated road, rail and air networks. Rates for public transport vary, being more expensive in the larger cities like Sydney and Auckland. Because of the size of Australia most people fly between major cities and pick up car rental within each state. Drive left.

Accommodation: A lot of premises calling themselves hotels are just plain bars. Older hotels may not have private toilets or bathrooms; breakfast may or may not be included. Check exactly what you are getting before you buy. Watch out for segregated bars in outback areas and bear in mind that outback accommodation is likely to be basic, even rough. Camping and hostel facilities are good. Bed and breakfast and other homestays - particularly on ranches and farms, are becoming increasingly popular, as are apartment/self-catering short-term rentals.

Background reading: Garner, Helen, *Postcards from Surfers* (Bloomsbury, 1989).

China

Climate

China is a vast country so the climate varies widely. The western part of the country has a lot of rain, with long warm summers in the south and short cool winters. Further north there is less rain, and the summers are not as long or as hot. In the east the mountains mean the climate varies with elevation and exposure.

China: tips

The events of Tiananmen Square on the night of 3 June, 1989 have caused a dilemma for those thinking of visiting China. Many travellers are reluctant to go in case their visit signifies approval of the way the government treats its people. On the other hand there are the Chinese people who appreciate the fact that people still want to visit their country and see for themselves what their lives are like.

*Despite the political tensions, visitors are safe and sexual harassment should not prove a problem for women.
*Be sensitive to the political climate.
*A black market whereby you can exchange your money for 'people's money' is lucrative, but risky.
*Never lose your temper in public. Public criticism of another person is considered cruel in the extreme.
*Tipping is considered an insult. Similarly gifts of any great value cause embarrassment. If the recipient refuses to take it, don't insist.
*Touching somebody is considered bad manners. A slight bow is appropriate when meeting somebody, or perhaps a handshake.
*Practise your chopsticks before you go; you are unlikely to be offered a knife or fork. When eating hold the bowl close to your mouth. Making a mess of the table is no shame. It is

China

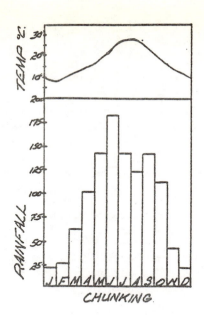

CHUNKING.

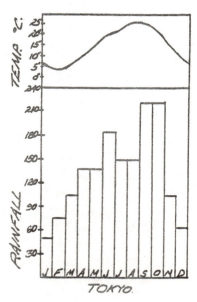

TOKYO.

polite to sample every dish offered and to leave some food on each serving plate. Don't start to eat or drink until invited by your host.

*If your group is welcomed into a building with applause the correct response is to clap back.

*If you rent a bicycle tie something distinctive on to it so that you will be able to pick it out from the thousands of others after parking.

Language: Chinese; dialects. English understood in cities.

Transport: Trains and buses are reasonably efficient but always jammed to capacity. Public transport is not expensive.

Accommodation: The Chinese International Travel Service arranges hotels for tourists - usually the most expensive. Cheaper rooms or dormitory accommodation can sometimes be had by visitors, if you are persistent. 15 per cent is usually added to the quoted price for accommodation, for heating in winter and air conditioning in summer.

Background reading: Croll, Elizabeth, *Chinese Women since Mao* (Zed Books, 1984).
Cuzak, Dympna, *Chinese Women Speak* (Century, 1985).

Europe

Climate

The climate is generally moderate, with only northern areas like Scandinavia and central European countries far from the coast suffering extreme cold. In Western Europe there are four distinct seasons (spring: February to April; summer: May to August; autumn: September to October; winter: November to January) with no guarantee of good weather in any of them. Winter is characterised by almost continuous cloud cover, with rain or sleet. In the Alps heavy snow showers alternate with brilliant sunshine while down south the Mediterranean is the ideal

Europe

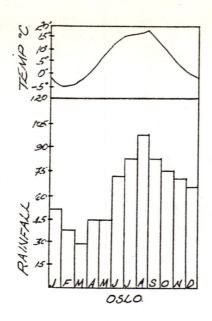

OSLO

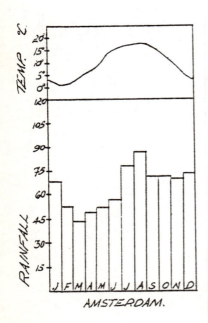

AMSTERDAM.

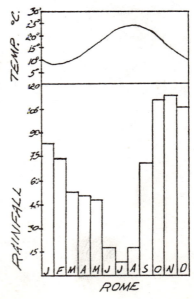

ROME

location for sun worshippers, hot for most of the year but rarely too hot or humid.

Europe: tips

*In rural areas you are unlikely to be bothered by men, except perhaps in bars. But in cities you will feel uneasy walking alone late at night. In Southern Europe - countries like Italy and Greece - street harassment is said to be more of a problem than in Scandinavia and other Northern countries, where the people see nothing noteworthy about women travelling, in groups, or alone.
*Europeans are usually tolerant of dress and behaviour, but it is best to avoid wearing shorts or skimpy tops into churches; check out beaches before going topless.
*Racism is rampant in Europe, and black, North African and Asian women often find themselves vulnerable to double discrimination on the basis of sex and race.
*Hitchhiking is illegal in parts of Europe.
*Use first names on invitation only. Those with academic titles expect you to use them. All business introductions, and indeed many personal ones, involve an exchange of business cards.
*Punctuality is essential, particularly in Northern European countries.

Language: More than fifty languages are spoken in Europe.

Transport: Costs vary depending on the extent to which transport is subsidised by the government. Transport services throughout north-western Europe are efficient, modern and expensive; in the south and east transport of all types tends to be older in design, but still regular and reasonably efficient, and a lot cheaper.

Accommodation: Western Europe is well serviced with hotel accommodation in all price categories, as well as excellent camping and hostelling facilities. In Eastern Europe hotels, even those in the luxury class, are basic (and cheap), although this looks set to change with the introduction of free market economies.

Recommended reading: Gadant, Monique, *Women of the Mediterranean* (Zed Books, 1986).

South East Asia

Climate

Spring (March and April) and autumn (September and October) are the best times for a visit to Japan or Korea. In the summer (June to August) Japanese cities are unpleasantly hot. Hong Kong has a subtropical climate: hot, humid and wet summer with cool, dry winter. September to early December is the best time to visit, when temperatures and humidity have fallen. Thailand, however, is best in November through to February as the wet season lasts from June to October, and March to May is extremely hot. The same is true of the Philippines. Malaysia and Singapore have no pronounced wet or dry seasons. October or April/May are usually the best times for a visit.

South-east Asia: tips

Much of this part of the world has a history of western colonisation so there is a mixture of west and east in the cultures.

*In Thailand it is considered very rude to show the soles of your feet. The foot is considered to be a low place, inhabited by low spirits. So never point your foot at a Thai. Be careful how you cross your legs and how you sit on the floor.
*Never touch a Thai or pat a Thai child on the head, as this is where the soul is held.
*The people of Indonesia are predominantly Muslim, but tend to be tolerant. Aggressive sexual harassment is rarely a problem. Indonesians seem not to understand the concept of a need for privacy or personal space so beloved of westerns. They might be personally insulted if you told them you

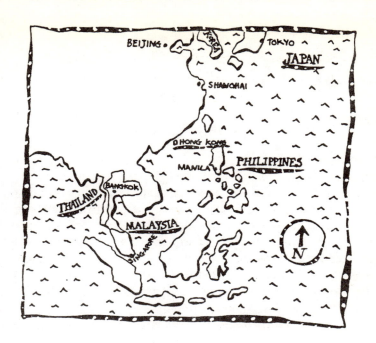

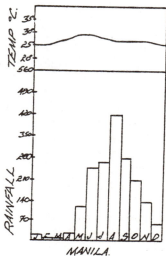

South East Asia

needed to be alone.
* If you are offered a gift in this part of the world, thank your benefactor, but wait for one or two more offers before accepting it. Receive the gift with both hands.
* In these countries the sex industry flourishes. Young Thai and Filipina women frequently face harassment from GIs and foreign tourists. This is upsetting and you must devise your own strategy for coping with it. Harassment rarely extends to the visitor; if it does it is usually in the form of catcalls or propositions.
* Japan's situation as the world's richest nation has led the Japanese to relegate visitors from less successful countries to second-class status. White westerners tend to be treated with traditional hospitality and courtesy but black or Asian visitors are discriminated against.
* Remove your shoes before entering a Japanese home.
* Japan has a well-deserved reputation for safety. Violence or harassment on the streets is rare. The culture is however deeply sexist. Pornography is everywhere, as is prostitution.

Transport: Varies from place to place. Excellent in Japan, if expensive. Generally slow and crowded but cheap. Each nation has it own (usually excellent) airline with domestic network and services between its base and other capitals. Driving conditions unsuitable for self-drive.

Accommodation: In Malaysia and Indonesia the best options are the cheap government rest-houses. Small hotels are often Chinese run: spartan, noisy but clean and efficient. It is worth noting that a single room has a double bed, while a double room (more expensive) has two singles. It is expected that you will haggle over the price of a room.

Background reading: Minority Rights Group, *Women In Asia* (MRG 1982).

Indian Subcontinent

Climate

Generally speaking the period from November to April is the best time to visit. Southern India is warm and humid all year round, while the northern Indian plains are extremely hot during the dry season. The South-West Monsoon brings rain from mid-May, starting in the south-west and spreading north and east. In Sri Lanka the Monsoons bring rain to the south-west from May to August and to the north-east from November to February. In Nepal rains begin in April, making March the best month for a visit.

India: tips

You will encounter a multitude of different attitudes to women in India because of the multi-racial make-up of the Indian people. But India is well used to lone women travellers.

*Harassment tends not to be a major problem, except in strongly Muslim areas, where going out alone may well invite verbal abuse and women are not expected to look men in the eye. Eve-teasing, the Indian term for groping, flourishes in crowded places, eg public transport, but such sexual violence against visitors is rare.
*Hustlers and beggars are everywhere in countless numbers. Every traveller has difficulty coming to terms with the poverty. If you give money, you'll find yourself surrounded by crowds. There is no feasible way you can give to them all. Expect it, and plan in advance how you will deal with it.
*Remove your shoes before entering an Indian home. Upon entering you may be adorned with a garland of flowers. This should be removed immediately as a sign of humility.
*Remove your shoes and don't carry leather goods into temples or mosques.
*Dress modestly particularly when visiting religious shrines.
*Mind how you eat. The majority of travellers to India become ill, with 'Delhi belly' if nothing else. Dysentry and hepatitis

Indian Subcontinent

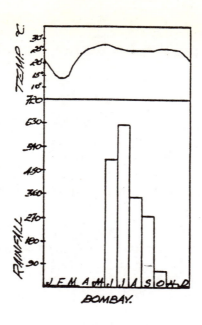

BOMBAY.

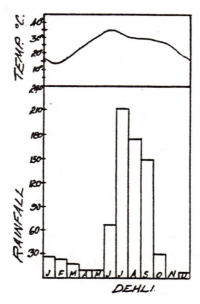

DEHLI.

are real risks. Be vigilant.
* Use auto-rickshaws rather than taxis; they are much cheaper.
* Theft is common.

Language: Hindi. English. Regional languages.

Transport: Indian railways are safe and comfortable, if slow. Advance booking is recommended. Buses range from dilapidated to sleek and modern. Bus travel within cities is overcrowded and unreliable, except perhaps in Bombay. Opt only for yellow-topped black taxis as these are licensed and have a fixed fare system.

Accommodation: All budgets and preferences are catered for. Converted maharaja's palaces give a fascinating glimpse of a vanished lifestyle. Be prepared to pay for accommodation in local currency once outside of the main population centres as travellers cheques and credit cards may not be acceptable.

Background reading: Kishwar, Madhu and Vanita, Ruth (eds), *In Search of Answers: Indian Women's Voices* (Zed Books, 1984).

Middle East

Climate

A large proportion of this area is desert. Iran and Iraq in the north suffer extremes of heat and cold with rain in the winter and spring. In the Mediterranean areas there are long sunny summers and mild wet winters. At the beginning and end of summer a dry, dusty, desert wind blows.

Middle East: tips

This is a very difficult part of the world for women travellers. In some countries women are allowed in to work, but they must stay in compounds and are subject to curfew. In others they are simply refused admittance. Even in Saudi Arabia, one of the more westernised countries, single women are now finding it

Middle East

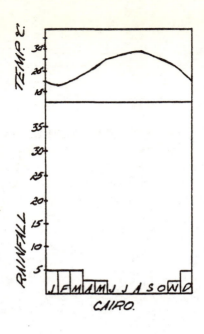

CAIRO.

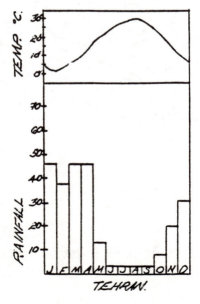

TEHRAN.

difficult to obtain a visa.

Getting in is just the first hurdle. Once there you may find hotels reluctant to register single women and you may be asked to produce a letter of permission to travel from your father or husband!

* Photographs to accompany your application to middle eastern countries must show you with your hair drawn back so that it is not visible and your head covered with a headscarf.
* Women exercising their right to wander freely are subjected to verbal and physical abuse - cars hooting and following them, staring and shouting, touching and jeering. But rape or physical attack in public is rare because punishment for crime is harsh.
* Travellers who have visited Israel will not be admitted to any of the Arab League countries. If your difficulty is explained to Israeli officials they will arrange to have entry notices and visas stamped on a separate piece of paper, rather than on your passport. Alternatively apply for a second passport.
* As usual in Islamic areas temper your behaviour and wear modest, conventional western clothes - tops with long sleeves and high necklines, loose fitting skirts and dresses and loose fitting light trousers, a headscarf. Leave shorts, short skirts, sun-tops or bikinis at home unless you will be in ex-patriot compounds where you can wear what you would at home. Religious police deal severely with women they consider to be immodestly dressed.
* Alcohol is prohibited in Islamic countries, except for Iraq.
* In Saudi Arabia it is illegal for women to drive and women venturing outside the special compounds which have been built for foreigners are subject to curfew.
* In Iran (if you can get in) you will be expected to cover your entire body and to hide your hair beneath a veil. Wandering freely in this country effectively brands you as a prostitute.
* Islamic religious custom demands that everything stops five times a day for prayer, which can be disruptive if you are engaged in business. As a foreigner you are not expected to kneel and face Mecca, but it is important that you respect the

religion and you must not display impatience or interrupt. During the religious feast of Ramadan no work is done after noon, another rule which you should respect.

Language: Arabic, Turkish, Kurdish, Farsi.

Transport: Services vary in quality, but all transport is cheap. Many areas have efficient buses and trains. Trains have women's compartments. In some places bus transport is segregated with women up front.

Accommodation: Plenty of reasonably priced accommodation. Some places will refuse to have a woman stay if she is travelling without a man.

Background reading: El Saadawi, Nawal, *The Hidden Face of Eve* (Zed Books, 1980).

Former Soviet Republics (C.I.S.)

Climate

Much of this area is beyond the Arctic Circle, with ice-bound coasts, permanent snow cover and tundra soils. The centre of the continent, the Steppes, is a vast low-lying plain, prone to severe and very changeable weather, with cold winters, searingly hot summers, strong winds and blizzards. The coastal areas are more comfortable than comparable latitudes inland, and in the south, particularly on the Black Sea, summers are warm and dry.

Former USSR: tips

Many westerners are only beginning to appreciate the widely contrasting cultures that make up this huge area - what used to be the largest country in the world - the USSR.

Until recently the usual method of travel to this area was a package deal arranged through Intourist, the main Soviet travel organisation. But new co-operatives are being formed which offer different travel options, and visitors today are allowed to

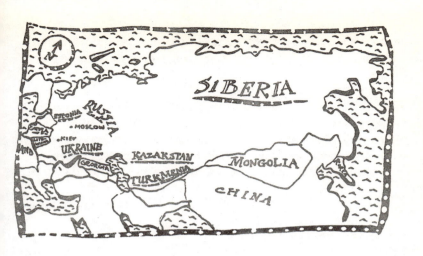

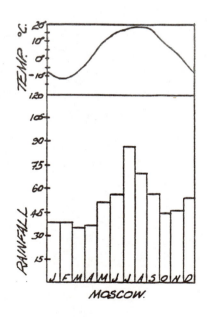

Former Soviet Republics (C.I.S.)

explore as they wish and even visit people's homes. Soviet citizens generally seem interested in meeting people from other countries and in discussing their own country's situation and future.

*Russia is just one country in this vast area. Never refer to the area as a whole as Russia, particularly in the republics.
*Plan your trip well in advance. Sorting out visas, contacts and travel details can take a long time. As things are changing rapidly make sure your information is up to date. Be prepared for a rapid change of itinerary due to unrest.
*It is illegal, and highly offensive to the citizens of any of these countries, to drop litter.
*Punctuality is important.
*Shaking hands and exchanging names is the normal procedure on introduction. Greetings among friends include hugging and kisses on the cheek.
*Travelling around any of these countries, alone or with a group, was traditionally safe. Sexual assault was rare in Russia and in the Baltic republics although in the Caucuses and eastern republics men occasionally labelled foreign women as 'loose' and made advances. Since the demise of the Soviet Union, however, pornography has become widely available so incidents of harassment and sexual violence are likely to increase.
*You will be approached for foreign currency and to exchange goods. It is safest to steer clear of the black market, as arrests have been made.

Language: Various.

Transport: Good transport network. Trains and buses tend to be old and slow, but are comfortable and cheap. Internal flights are cheap.

Accommodation: The five-star hotel business is just beginning to move into this area. In the meantime there is plenty of cheap, basic accommodation to be had. There are also good camping facilities throughout the country and these Intourist camp-sites also have wooden huts.

Background reading: Mamonova, Tatavana (ed), *Writings from The Soviet Union* (Blackwell, 1984).

South and Central America

Climate

Most of South America has a tropical or subtropical climate, warm or hot all year round. In Central America the area facing the Caribbean sea has heavy rainfall throughout the year, at its worst from October to February. The rest has a dry winter season from November to April.

The west side of the South American continent is dominated by the Andes mountains, where temperatures are affected by altitude. Rain is usually confined to the summer months both north and south of the equator, except in South Brazil, south-west Chile (the coastal area) and the eastern coast of Argentina and Uruguay. The Amazon basin and the coastlands of Colombia and Ecuador also receive a lot of rainfall.

June to October is the best time to visit Argentina and Brazil, Paraguay and Ecuador. Much of Chile has a dry, desert climate but Bolivia, just over the border, experiences heavy rains from May to November in the mountains and throughout the year in the rest of the country.

*Machismo prevails throughout Central and South America. Tales of the promiscuity of foreign women abound and are compounded by resentment of foreign women's apparent wealth. Self-confidence and some Spanish help. The best ploy seems to be to ignore it.
*Being a foreigner you are by definition wealthy in the eyes of most people in South America. Theft is common with jewellery the favourite target, after money. Remove anything valuable before you go out and don't flaunt money.

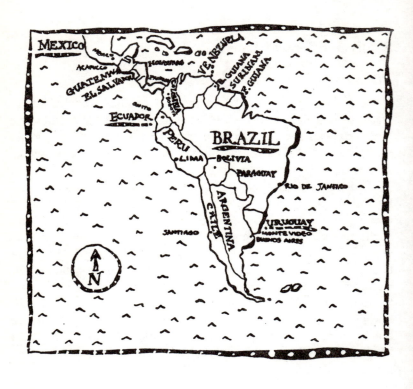

South & Central America

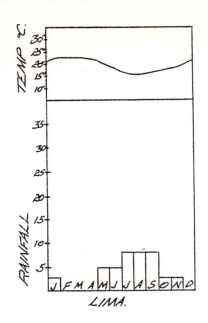

LIMA.

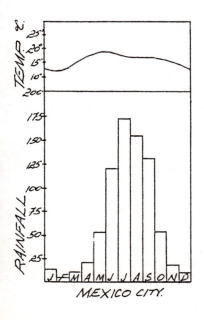

MEXICO CITY.

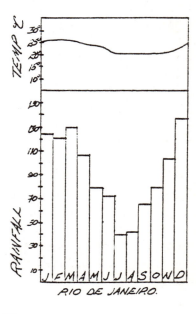

RIO DE JANEIRO.

Thieves may work in family groups, using their appealing children to catch you off-guard.
* Be wary of the police. Their wages are low and bribery is seen as a perk of the job.
* The drug business is big business in South America. Buses and other forms of transport are often raided by police and it is not unusual for drug pushers to set foreigners up. Never carry a package for anybody else without first checking the contents.
* In general in Central America compliments on personality are more appreciated than compliments on possessions. If you admire something which somebody owns they are likely to offer it to you. If you compliment a personal quality, however, they will be pleased.
* Be careful when asking directions. Most Latin Americans will give you the wrong answer rather than say they don't know. Don't ask leading questions.
* Many Latin American countries will not let you in unless you have an outward bound ticket.
* Uninvited visitors in Costa Rica can be refused admittance.
* In El Salvador, do not be alarmed to find a local put his or her arm around you. This is a sign of friendship, no more.
* Bolivia and Chile are probably the two safest South American countries for women travellers. A woman alone may attract attention but it is unlikely to be threatening. Colombia, Peru and Brazil are probably the most wearing in terms of crime and sexual harassment.
* Hitching is dangerous in South American countries.

Language: Spanish. Portuguese. Various local languages.

Transport: Transport systems are improving but are still poor in many areas. Trains tend to be slower and less comfortable than buses. Theft of cars is common. Apply at least two anti-theft devices and remove detachable items like wipers or mirrors. It is cheaper to buy internal flights as you go rather than in advance before you leave home.

Accommodation: Plenty of hotels, most cheap and fairly basic. There are hostels in Argentina, Brazil, Mexico and Uruguay and a growing network of organised camp-sites.

Background reading: Latin American and Caribbean Women's Collective, *Slaves of Slaves: The Challenge of Latin American Women* (Zed Books, 1982).

The USA and Canada

Climate

Almost half of Canada and North Alaska is beyond the Arctic Circle. The coast is ice-bound, the northern areas have permanent snow cover and the ground is tundra.

The centre of the North American continent is a vast low-lying plain, prone to severe and very changeable weather. Inland states can experience cold winters, searingly hot summers, strong winds and blizzards. The east coast tends to be wet while the west coast is dry.

The coastal areas are more comfortable and are the areas which see the largest population densities. The west coast is protected from winds by the Rockies, and sea breezes keep things cool further south. North-east coasts suffer extremes of temperature while down south weather is warm all year round. The tropical south-east is vulnerable to hurricanes and tornadoes.

USA and Canada: tips.

*So many women travel in these regions that you won't stand out, but hesitant visitors are fair game for thieves and muggers in big cities. Act confidently, try not to look as if you are lost, and don't consult your map in the street. Avoid carrying more cash than you need for the day.

*Americans generally eat earlier than Europeans and in many areas the restaurant kitchens are closed by 9.00pm, though the proliferation of fast food joints usually means you will

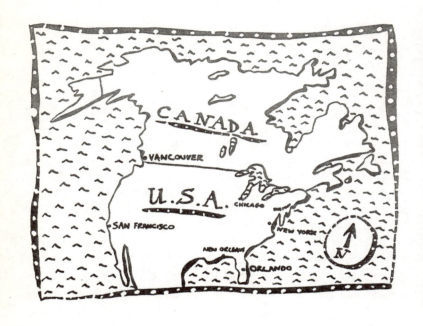

USA & Canada

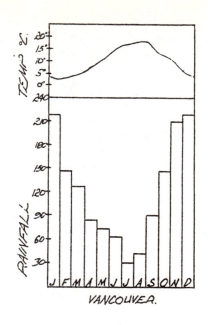
VANCOUVER.

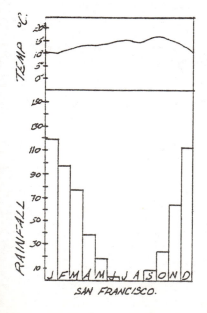
SAN FRANCISCO.

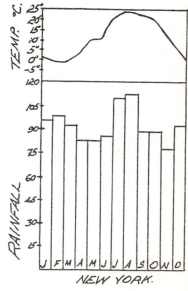
NEW YORK.

be able to get something to eat at most times of the day.
* Breakfast business meetings are popular. Be punctual for business meetings; attend fifteen minutes late for an invitation to someone's home.
* Americans don't stand on ceremony. Men usually shake hands during introductions, women not so often. After introductions, it will generally be first name terms all round.
* Each US state has its own legislation, religious and political affiliation and attitudes to women.
* Canadians are perceived to be more conservative than Americans.
* Don't think about going to North America without an insurance policy that provides huge medical cover - at least £1,000,000. If you are ill or injured and you cannot prove that you can pay for treatment, you may not be able to get any.

Language: English. French (Quebec, Canada).

Transport: Buses, trains, air travel and car hire are all widely available, and of a good standard - but be prepared to pay for car-hire, or more than standard class travel.

Accommodation: All standards and prices available widely. Some major cities have women-only hotels. Excellent hostelling and camping facilities. on the outskirts of many cities.

Background reading: Steinem, Gloria, *Outrageous Acts and Everyday Rebellions* (Flamingo 1985).

Kome, Peney, *Women of Influence: Canadian Women and Politics* (Doubleday 1985)

Getting the paperwork out of the way

All travellers need documents and exactly what you require depends on your nationality and status, and the purpose of your visit - tourist or business. What you require in the way of documents may change dramatically, even overnight, depending on the political situation of the country you are visiting.

This section deals with paperwork like passports, visas, insurance and money. If possible read it well in advance of your departure, as some paperwork may take time to organise.

All documents should be photocopied and kept in a separate place from the originals, in case of theft.

Passport

Every woman should have her own valid passport. It confirms your identity, nationality and status and you won't get far without it. You will be asked to produce it at the border of every country you visit. It may also be requested by officials on many occasions during your travels - when changing money, for example, or checking into accommodation. Leave a photocopy of it at home, in case of theft, and carry a copy with you.

Each country has a central passport office, run by the government of that country, and that's where you apply (in writing) for a passport. The general requirements for a passport to be issued are your birth certificate, two passport photographs and a referee who will sign the application form and endorse the back of one of your photographs to confirm it is you. Further information is available from your passport office. See local telephone directory for details.

It is usual for children up to the age of sixteen to be included on their parents', or guardians', passports. A dependent can only travel with the parent who has been named, so if you are a two-parent family it is a good idea to enter them on the passport of both parents. Even if you always travel together an emergency could require the 'unnamed' parent to travel with the children, providing you with a bureaucratic obstacle you could do without.

Children between five and sixteen years may hold passports of their own. This is convenient if they are to travel without a member of their family, on school trips and such like. In exceptional cases this may be extended to children under five years.

Visas

You don't need a visa for every country; it depends on your nationality and where you wish to go. However, visa requirements change rapidly as the political situation within a country changes. You therefore need to check, well in advance of your trip, whether or not you need to apply for a visa. Ask your travel agent or contact the consulate/embassy of the country concerned to get up-to-date information.

The procedure for obtaining a visa is usually as follows: apply to the consulate or embassy of the country you wish to visit which is based in your home country, or, if they have no representation, to one of its overseas offices. Your travel agent will advise you.

Applications are made in writing, by filling out a form and returning it to the embassy or consulate with a fee and your passport. You should send your application by recorded delivery, and enclose a stamped addressed envelope to make sure your passport gets back safely, in time. Always allow several weeks for this. Applying in person by visiting the consulate can speed up the process, but often involves long queuing.

Visa applications can sometimes be complicated and it is possible to hand the job over to a visa service, who, for a fee, can manage the entire process for you, saving time and hassle. Consult your local telephone book for visa services near you.

Visas are issued for a defined length of time, usually, but not always, for a period of between thirty and ninety days. If you are in doubt about your length of stay, find out in advance whether it will be possible to renew your visa while in the country and what the procedure there is.

The purpose of your visit may be another restriction. Some countries, eg Saudi Arabia or Albania at time of writing, discourage tourists and casual visitors and allow people in for business purposes only. If you are travelling on a business visa, expect to be asked to produce a company letter stating the nature of your business, and the address of your contacts within the country.

You may need to produce evidence that you have sufficient funds to support yourself on your travels. You may be asked for a letter from your employer confirming that you will be returning to employment after your trip. You may also be asked to show certificates of vaccination.

For a visa application produce photos of yourself looking neat and tidy, and as conventional as possible, and carry this image through when visiting consulates, or entering a country. People have been refused a visa on the grounds of appearance.

If you intend to do a lot of coming and going between countries, check whether you can apply from the outset for a multiple entry visa, which will save separate application every time you want to re-enter the country.

For visa information check with:
  *a good travel agent
  *the relevant consulate
  *a visa service

Insurance

Adequate insurance is vital for every trip, no matter how short. It does not usually cost a lot, and you really never do know what might happen when you are travelling.

For most people, the decision is not whether to take out insurance, but which policy to go for.

Before buying a policy consider the following points:

*Don't take the first scheme you are offered without reading the small print. Even if it does look all right, don't accept it until you have shopped around to see if you can get equivalent cover for less money. Some agencies may insist that you take the policy they offer; argue this one with them.
*Does the policy give adequate medical cover, particularly if you are visiting the US, where medical costs are astronomical?
*Is the cost of medical treatment paid by your insurer directly to the hospital or medical centre (preferable, particularly if you are travelling far away, or to somewhere remote), or do you have to pay the bill, and wait to be reimbursed later?
*What arrangements are there for you to make claims during your travels? Does the policy include a 24-hour emergency service, providing immediate assistance in a language you can understand, wherever you are?
*What happens if a companion has an accident - are you covered to return home if you so wish?
*Are you covered for dangerous sports if your trip includes skiing, mountaineering, scuba diving?
*If you anticipate that you might take out a motor scooter to travel around, does your policy cover you?
*What about baggage cover? Do you need the cover offered, or is it adding unnecessary expense to your policy?
*Are you pregnant? Often policies have exclusions for pregnant women.
*Can you extend cover if you are away longer than planned? At what cost?

If you are joining a pre-arranged tour, or package holiday, you will find that having adequate insurance cover is compulsory. Some tour operators prefer you to use their own schemes; if you choose not to do this they will want to see a copy of any policy you take out independently to check you have adequate cover.

Money options

For the travelling woman of the '90s there are ever more options on how to carry money. Deciding which of those options to go for can be fairly complicated if you want the best possible financial deal, particularly for travellers within European countries who have the widest number of choices.

Cash

Most travellers bring some local currency in cash with them - at least enough for bus/taxi fares from the airport or station, and a cup of coffee or a snack. The currency you require may not always be available in your home country, however, in which case you will find yourself queuing at the airport bank when you arrive, thus paying higher rates of exchange than necessary and having to queue longer for buses/taxis.

*Ask your bank whether they can order currency for you, and if it is an unusual currency, or a country where supply of currency is erratic, allow several weeks for this.
*Try to obtain as many small denomination notes as possible. Change is often a problem, particularly in Third World countries.
*If the country you are visiting has a weak currency you might also want to take an emergency supply of cash in a strong currency, eg US dollars or Deutchmarks.

Travellers' cheques

The advantage of travellers' cheques over bank notes is that they are replaceable if they are stolen or lost. Suppliers have a network of overseas offices, agents and sub-agents who will replace cheques for you.

* Travellers' cheques are now available in a range of foreign currencies, which means you may have the choice of buying cheques in your home currency and changing them into local currency as you need, or carrying cheques in local currency and being able to use them as you would bank notes to settle bills, without having to convert them.
* When cashing travellers' cheques, shop around. Exchanges may offer slightly differing rates, or take differing amounts of commission. Decide on the denominations you wish to bring with you and as you cash them, remember the charge for each transaction.
* Once you have bought your cheques, sign each one immediately as no refund will be given for unsigned cheques. Most banks will ask you to do this before you leave the counter. You then countersign each cheque as you cash it, in front of the person accepting it.
* Keep a list of your cheque numbers separately and mark them off as you spend them, so that you have a record of what's missing if you lose them, or if they are stolen. As with cash, it is useful to take a mixture of large and small denominations.
* In Third World countries take care in choosing your supplier of travellers' cheques, as the network of agents and offices may not be very widespread. Suppliers have lists of their agents abroad, so you should shop around to see which is the nearest to you, and compare the services offered before you buy your cheques.

Plastic money - credit cards and charge cards

Such cards can be a useful back-up to other money resources, enabling you to settle some bills without cash and to obtain some items on credit.

* If you are on the move and will not have a permanent address where monthly statements can be sent, arrange for your home bank to pay your credit/charge card company for you.
* Many credit card companies provide accident insurance if a travel ticket is booked using their card. Many cards are also covered by credit legislation which means that if you pay for your travel arrangements by credit card and the company or airline goes bankrupt, you may be able to get your money back through your credit card company.
* Charge cards like American Express work differently. You can't obtain emergency cash, and insurance cover is different to that offered by credit card companies. There is a membership fee, and annual charge, and they offer no credit terms. For the average traveller the only advantage of a charge card is that it is a safer alternative than cash.
* For the business traveller, however, charge cards do offer a range of extra services, like immediate spending money if luggage is lost or delayed, guaranteed hotel reservations and special airport departure lounges. They also bestow a certain prestige, something which a business woman may find useful.

Eurocheques

As the name implies, this scheme is limited mainly to Europe, although a few banks in capital cities elsewhere do participate. You draw money as you use it directly from your home account using a special Eurocard and Eurocheque in exactly the same way that you would at home. You can use cheques to settle bills or to draw cash from a bank. You can also take out money from cash dispensers.

To obtain a Eurocard you must apply in advance for your card and set of cheques through your own bank, at home. Allow several weeks for this. You should receive a list of cash dispenser locations in the area to which you will travel. If your travels will take you outside Europe, ask for a list of other participants too.

Getting cash from home

OK, you hope it will never happen, but you just might need to call on home to forward money in an emergency. Have your contingency plans made before you go. Discuss with your bank manager, and somebody else at home, what you will do if you run out of cash, or lose all your possessions. Make sure that both you and the person at home who is responsible knows what procedures to follow. Carry any necessary addresses or telephone numbers with you at all times, separate from your other documents.

*Most European, North American and Far Eastern banks have methods of transferring money in an emergency, eg. the SWIFT (Society for World-wide Interbank Financial Telecommunications) system, or telex transfers, or foreign drafts. Your bank manager will have details.
*International Money Orders is another option available from some banks.
*If you are in one of the countries covered by Eurocheque and your sender is a Eurocheque holder, she can send you a Eurocheque.

Driving

An international driving licence should be applied for well in advance of leaving home. Once you have a full driving licence, this should be a mere formality. Check restrictions/obligations in countries to which you are travelling.

Money and paperwork check-list

*Passport
*Visa (if necessary)

*Insurance policy
*Local currency (don't forget your money belt)
*Travellers'/Euro Cheques
*'Plastic' money
*Driver's licence

Staying healthy and well

It is essential that any woman setting off to travel should feel confident that she will return home healthy and well, rested and relaxed.

Often our behaviour on holiday - notably a determination to have a good time - can have consequences for our health and safety. Accidents are a frequent cause of death among travellers; upset tummy is common, often brought on by over-indulgence; alcoholic poisoning is another common problem.

Then there is the fact that if an injury or illness does occur while you are abroad, you are at a disadvantage - communication with physicians or hospital staff will be difficult if you are unfamiliar with the language. Even if you know the language the local medical system is likely to be different to that in your own country and you may find yourself unable to pay for emergency treatment. In addition, medical standards may not be what you are used to.

Unfortunately few travel agents give their clients good health information. And few travel brochures make any useful reference to health. Little wonder, then, that surveys have shown that few travellers are informed of potential health hazards when travelling.

The following section outlines, in alphabetical order, the major threats to your health and precautions and remedies for you to take. If you are travelling only within North America, Northern Europe and Australia, infectious or parasitic diseases will be largely irrelevant. But if you are travelling elsewhere, particularly in Africa, Asia or Latin America, you should pay particular attention to these sections and to the section on food hygiene.

Accidents

More travellers die from accidents than any other cause and most accidents are avoidable. When travelling resist any temptation to expose yourself to risks which you would not take in your home environment.

Many women are nervous of flying but it is, in fact, one of the safest forms of travel. Do listen to the safety pep-talk which you are given at the start of your flight. Travelling in a private plane is more dangerous. Be aware of the risks you run and decide whether they are justified.

Statistically, motoring abroad is far more dangerous than flying. Regardless of the safety regulations of the country in which you are travelling you should always wear a seat-belt. Do not drive if you are tired, or if you have indulged in alcohol.

Moped accidents in holiday resorts are common and aggravated by the fact that few people who rent mopeds wear crash helmets or protective clothing. If you do rent a moped, be stringent about adhering to the rules of the road, particularly if you are driving on the 'wrong' side of the road. Wear protective gear and be careful.

If travelling by boat locate safety devices such as life-jackets, muster stations for life-boats and emergency exits, to be prepared for the unlikely event of an accident.

Rail and coach accidents are rare and there is little or nothing you can do to prevent them, but do use reputable companies.

In your hotel or accommodation, identify fire extinguishers, emergency exits and any other safety features. Keep a torch handy in your room. If a lift looks unsafe use the stairs instead. Hotel and apartment balconies are notoriously unsafe. Keep children away from them and avoid the area yourself if you have the slightest doubt about its safety.

If you drink alcohol be aware that it increases your vulnerability to injury. Do not swim or drink after alcohol or drug consumption.

Wherever you are have a strategy for safety, knowing how

you should behave in the event of an emergency and what your escape route should be.

Aids

On the increase the world over, Aids (Acquired Immunodeficiency Syndrome) is a specific defect in the human body's immune (defence) system, leaving it vulnerable to diseases and illnesses that would not otherwise affect it. Aids is caused by infection with a virus called human immunodeficiency virus (HIV). Aids has a high fatality rate.

The virus is transmitted through sexual contact with an infected person or via contaminated blood that comes into contact with your own. There is no evidence to show that mosquitoes are capable of transmitting this virus. There is also danger of infection from blood transfusions in some countries.

The majority of those infected with HIV will have no obvious symptoms, but are capable of transmitting the disease. If you have sex with a person whose sexual history is unknown to you, protect yourself: insist on a condom and spermicide.

Be wary of injections abroad, remembering that HIV can be transmitted by contaminated needles. Take your own disposable needles (see Medical kit at the end of this chapter). In an emergency, do not take chances.

Avoid any skin piercing procedures - tattooing, ear piercing, acupuncture - unless you are certain that new or sterile equipment is being used.

There is no cure for Aids or HIV at present.

Bites and stings

Dog bites are most common, but women have been bitten or attacked by cattle, rams and goats, pigs and cats. In tropical or safari areas the prospect of attack from wild animals should be noted and safety precautions taken.

In the event of a bite, the affected area should be immediately and thoroughly cleaned. Wash with soap or detergent and running water for at least five minutes, removing any visible dirt or other foreign material. Apply a virucidal agent such as iodine. In an emergency strong alcohol like whiskey or gin will do.

All bites carry a heavy risk of infection and a serious attack requires immediate hospital attention, to take precautions against infection, tetanus and rabies. Rabies causes at least 50,000 human deaths each year, the most common cause being a bite from a rabid dog.

If you are travelling in a rabies endemic area avoid close contact with mammals. If you are bitten while travelling in such an area you should inform your doctor - giving details of when and where the bite occurred, its severity and the species, appearance and behaviour of the biting animal. On the basis of this information he or she will decide whether a rabies vaccination is necessary.

Snakebites/jellyfish stings can also be fatal. Those unfortunate enough to be bitten by a snake should have the bitten limb immobilised with a splint or sling, and immediately be taken to hospital. Aspirin should not be given for pain; instead give paracetomol. If the snake has been killed take it to the hospital too, but do not handle it with bare hands.

Scorpion stings produce venom which may damage the heart, causing death, particularly among children. Local pain is severe; best treated with a local anaesthetic. Visit a doctor immediately upon being stung.

Fish stings are excruciatingly painful and can also cause vomiting and diarrhoea. To relieve the intense pain immerse the stung limb in very hot water. Visit a doctor for a local anaesthetic.

Stings from bees and wasps are less painful, unless an allergy develops. Embedded stings should be scraped out, not pinched. Aspirin may be used for pain. Those allergic to stings should carry an identifying tag.

To avoid discomfort and possible allergies resulting from

mosquito bites, do carry a travel mosquito killer pack, with enough foil-wrapped tablets to cater for your trip.

Constipation

A less common problem than diarrhoea, constipation can result from dehydration or dietary changes. A high fluid intake and high fibre diet are preferable to medication.

Contraception

It's best to bring your own contraception with you. If your luggage gets stolen, or if you encounter a problem with the method you are using, local hospitals or family planning agencies should be able to help.

The pill

When travelling you must continue to take the pill regularly without leaving more than twenty hours between doses. Be particularly vigilant when crossing time zones. The progesterone-only mini-pill has to be taken at the same time to work properly, which makes travel across time zones difficult. Check with your doctor.

If you are vomiting or have diarrhoea the pill probably won't be absorbed efficiently. If diarrhoea sets in less than twelve hours after taking the pill, consider yourself unprotected. Keep on taking your pills but use additional contraceptive measures for fourteen days after you recover.

Should you vomit less than three hours after taking a pill, you will still be protected if you can manage to take, and keep down, another pill within twelve hours in the case of the combined pill and three hours in the case of the mini-pill. If this is successful, go back to taking your pills at the normal time. If you carry on being sick, consider yourself unprotected, and, as above, continue taking your pill at the normal time but use extra contraceptive measures for fourteen days after you recover.

Some antibiotics have been shown to reduce the effectiveness of the combined pill. Rifampicin, which is prescribed for tuberculosis, is one. There are also suspicions about ampicillin and tetracycline. Bear this in mind when deciding on an antibiotic. Antibiotics do not affect the efficiency of the mini-pill.

The sheath, diaphragm, spermicides and IUD

Condoms and diaphragms may deteriorate in tropical climates, becoming thin and sticky. Check carefully before use. Keep them as cool as possible in a closed container.

Spermicides are also affected by the heat and direct sunlight. Pessaries, foaming tablets and any other spermicides which work by being dissolved are particularly vulnerable, and may melt in the heat. Instead bring jellies or creams and keep them refrigerated if possible.

If problems are going to occur with an IUD - expulsion, pain or severe bleeding - they are most likely to happen in the first month after the coil has been fitted. If you are going to have an IUD inserted, do so well in advance of your trip. If troubles occur en route don't ignore them. Go to a doctor for a check-up.

Cholera

Cholera is a severe bowel infection causing fluid loss which can be fatal within hours. The disease is associated with poor sanitation, passed on by faecal contamination of food.

The cholera vaccine which has been widely used for some years offers poor protection; estimates of its effectiveness vary between 40 per cent and 80 per cent. Only in areas of recent outbreak will you require a certificate of vaccination (check before you travel), although it may be worth taking the vaccine if you are going to a part of the world where the disease is prevalent (many countries in Asia, Africa and the Middle East).

A single dose of cholera vaccine provides certification, valid for six months, starting six days after vaccination.

Cystitis

Symptoms of cystitis are frequent and uncomfortable urination, usually due to contamination of the urinary passage with bacteria from the sufferer's anal area. The best treatment is plenty of fluid, particularly orange juice or the juice of other citrus fruits which alter the acidity of the urine, providing relief from painful symptoms.

Food hygiene

Food should never be assumed to be safe unless it is known to have been freshly and thoroughly cooked. No red colour should remain in meat. Shellfish should be boiled vigorously for at least ten minutes.

Fruit and vegetables should be freshly cooked or freshly peeled. Don't eat uncooked food or raw salads in regions where heat and unhygienic conditions prevail. Buy fruit from the market, wash it in purified water and peel it yourself.

Eat only hot food, which has just been cooked. Tepid food, especially if it is uncovered or being kept warm, is a breeding ground for bacteria. Also avoid food that looks as if it has been handled since cooking.

Familiarise yourself with food importation laws governing any country you wish to enter carrying packed food from home.

Assume all water and ice is unsafe unless you positively know otherwise. Always keep a water bottle full of boiled, purified water. If that is not possible, boil water for at least ten minutes. If you are still unsure about it add purifying tablets as an extra precaution.

Don't use water for adding to juice or washing children's utensils unless you know it has been boiled. Don't forget teeth washing. Clean your teeth and wash toothbrushes in boiled/purified water.

Be scrupulous about washing your hands after visiting the lavatory and before you eat. Carry wet wipes for the times when no hand basins are available.

Hospitality can be awkward. It takes great determination to refuse food prepared unhygienically by someone who has gone to great lengths to please an honoured visitor. Do not relax your standards: plead illness, toy with the food, but do not eat it if you suspect it to have been unhygienically prepared. Real offence is seldom taken.

Heat exhaustion

If you do not drink enough liquids or eat enough salt to replace what your body is losing through sweat you can experience nausea, headaches, tiredness and light-headedness. To treat, lie down somewhere cool and drink plenty of fruit juice or slightly salted water.

Hepatitis

There are at least six different forms of hepatitis. The viruses responsible for Hepatitis A and B are well understood, those responsible for Hepatitis D, or non-A non-B, less so.

The symptoms of all forms of hepatitis are similar - acute inflammation of the liver leading to fever and chills, headaches and other aches and pains, fatigue and weakness of the limbs. Later nausea and vomiting are followed by dark urine and jaundice of the skin and whites of the eyes. Complete liver failure may set in and the sufferer may go into a coma.

Hepatitis A is spread by faecal contamination of drinking water and food, so it is common where hygienic and sanitary conditions are poor. When visiting endemic areas it is recommended to have an injection of gamma-globulin as a protection against Hepatitis A and to observe strict hygiene.

Hepatitis B is passed in similar ways to Aids - through blood serum, non-sterile needles and sexual contact with infected people - so similar precautions apply.

There is no specific treatment for hepatitis. Bed rest is recommended during the acute phase. Sufferers are advised to follow a low-fat diet and to abstain from drinking alcohol.

Malaria

Malaria is a parasitic disease spread by mosquitoes. A year may elapse before symptoms of the disease appear, particularly if anti-malarial drugs have been taken. The symptoms are fever and chills with sweating, headache and general malaise. Jaundice may develop. Because of the delay in developing the disease, and because the early symptoms are non-specific, the condition may not be diagnosed.

Malaria is widespread in the tropical areas of the world. The treatment is with drugs - chloroquine, quinine or primaquine, depending on the type of malaria. Prevention is achieved by avoiding mosquito bites; wear insect repellent, use a mosquito net over your bed and/or buy one of the many preventive devices on the market.

Drugs can prevent malaria but they need to be taken one to two weeks before entering a malarial area and continued for six to eight weeks afterwards. Consult your doctor for advice. Pregnant women should avoid malarious areas.

Mountain sickness

The higher the body goes, the less easy it is for it to absorb the oxygen it needs into the bloodstream. Anybody moving rapidly to an altitude of 10,000 feet or more may experience breathlessness. Acute mountain sickness may occur if you go above 15,000 feet, leading to headache, nausea and weakness.

The body compensates by producing more red blood cells but this takes time, so acclimatise yourself slowly by making a gradual ascent.

Anyone developing symptoms like blue lips, a phlegmy cough, persistent headache or lack of urine should return to a lower altitude *immediately*, as occasionally mountain sickness can intensify into an oedema.

Poliomyelitis

Largely eliminated in developed countries polio remains common in the Third World. It is caused by a virus and results in paralysis of the muscles, particularly those of the limbs.

The polio vaccine is very effective and if you are travelling anywhere outside Europe, North America, Canada, Australia or New Zealand you should have this vaccine. Those who have never been immunised should receive three doses of polio vaccine at monthly intervals. Those who have been fully immunised in the past should have a booster every ten years if travelling to problem areas.

Prickly heat

A rash of red spots may occur if sweat is unable to evaporate from the skin and the ducts of the sweat glands get blocked. This is called prickly heat. To prevent it keep cool: wear loose-fitting clothing, preferably cotton. If you get a bad case avoid any strenuous activity which makes you sweat. Wash often with cold water and without soap. Calamine lotion offers relief.

Sexually transmitted diseases (STDs)

The only guarantee of avoiding an STD is abstinence from sex or sex only with a partner who is known to be disease free and absolutely faithful. Sexual contact with those who have had a lot of other partners is more risky than with those who had few, though knowing that is unlikely to make any difference if passion strikes!

Condoms, used properly (kept intact and kept on from start to finish) protect against STDs, as do caps and spermicide. Avoid unprotected sex with a partner unknown to you.

Common forms of STD include chlamydia which can cause pelvic inflammation and infertility; gonorrhea, which has similar symptoms; herpes, a disease which often recurs and which may be linked to cancer of the cervix; syphilis, which can lead to damage of the nervous system.

Signs of STDs include abnormal vaginal discharge, soreness, small ulcers or warty growths. If you suspect you have contracted an STD keep a record of symptoms and get a medical examination as soon as possible. If a disease is treated early, complications can be avoided.

Thrush

Vaginal thrush produces a thick white discharge which looks rather like cottage cheese and makes the vulva extremely itchy. The yeasts which cause it are always present in the body, but during a thrush infection they are temporarily flourishing.

There may be a variety of reasons for this. For some women being on the pill may produce the right environment for the infection to grow. Antibiotics, by killing off other organisms in the body, can encourage them to thrive. They favour warm, wet conditions, and a sugary environment.

To minimise the risk of developing thrush, wear cotton underwear and loosely fitting trousers or skirts. Always wipe and wash between your legs from front to back, to prevent yeasts from the bowel moving into the vulva.

Nystatin pessaries, clotrimazole cream, or gentian violet normally clear an infection. There are other methods which can work, particularly if you start them at the first signs of trouble.

Plain yoghurt contains bacteria which fights the yeasts. Eat lots of it, while cutting down on carbohydrates and cutting out alcohol. Rubbing yoghurt on to the infected area will soothe the itching. Use a tampon dipped in yoghurt for the vagina.

A bath or bidet laced with vinegar is also a help. Again you can soak a tampon in a solution of vinegar and water.

If you are prone to thrush, and you have a favourite remedy, take it with you.

Travel sickness

Often termed 'motion sickness' it is characterised by nausea and vomiting accompanied by cold sweating. Travel sickness can be induced by sea, air or car travel, occurring whenever we are exposed to unfamiliar motion. The degree of sickness is affected by the intensity and duration of the trip and the susceptibility of the individual. Susceptibility decreases with age but women are more susceptible than men of the same age.

If space permits, it helps to lie down. Symptoms are also decreased by closing the eyes, or keeping the eye fixed on the horizon or other fixed point outside the vehicle. Reading or writing aggravate symptoms.

There are a number of drug treatments on the market but none completely prevents symptoms in everyone in all environments. Consult your doctor if you are highly susceptible.

Typhoid fever

Typhoid is caused by bacteria called salmonella typhi. Early symptoms are headache, sore throat and fever, followed by abdominal pain, constipation and diarrhoea. Typhoid is a serious illness and can be fatal. An effective injected vaccine is available and should be taken if you are travelling to areas where sanitation is poor.

Tetanus

Tetanus is caused by infection of an open wound leading to impairment of the nervous system. Continuous muscle contraction and spasms and respiratory problems are the main symptoms.

The best protection is immunisation in infancy, with booster doses every ten years. It is safe to have tetanus injections at the same time as other immunisations. Another important preventive measure is the proper cleaning of wounds.

Vaccination

Depending on your destination you need to plan your immunisation programme some months before you go abroad, as a full immunisation schedule may take two months or more to complete. If you have less time, though, don't think there is no point in going for vaccinations. Valuable protection can still be gained.

There are now very few mandatory immunisation requirements, but that does not mean that they are no longer necessary. Vaccination regulations are designed to protect countries rather than travellers. It is your own responsibility to ensure that you are immunised if visiting countries outside the US, Canada, Northern Europe, Australia or New Zealand.

Your travel agent is likely to inform you about compulsory vaccinations for diseases like yellow fever or cholera. But you must consult a doctor or immunisation centre for more detailed information.

Yellow fever

A disease of monkeys living in tropical rain forests, yellow fever is spread by mosquitoes. The infection can range from mild to life threatening. Fever, headache and abdominal pain is followed by liver and kidney failure. Treatments aim to relieve symptoms and prevent complications from setting in. A single injection is highly effective, granting protection for a minimum period of ten years. If you are visiting Africa or South America you should ensure that you are vaccinated.

Many countries in the yellow fever zones of these countries require a certificate of vaccination. Don't forget your plug-in mosquito killer pack and tablets.

Medical kit

Everybody has different medical requirements. The list below gives the items which you might have most cause to use. Add to these any medication which you must take regularly - don't forget the contraceptive pill - and medication for any illnesses to which you are particularly prone.

A sandwich box, or any other plastic container with a close fitting lid, is an excellent container for your medical supplies, keeping everything together and also providing additional protection against spillage. A sealed plastic bag is best for dressings, sachets or anything in powdered form.

Label your medicines clearly, saying what they are, and giving directions for their use. Cover the label with plastic or sellotape to ensure that the directions remain legible.

Medicines are known by two names, their generic names, usually the same all over the world, and their brand names, which will vary. If you have to buy medicine abroad, make sure you get what you want.

Medical kit check-list

*Pain-killers
*Anti-diarrhoea medication
*Anti-fungal medication
*Antiseptic

* Antihistamines for stings or bites
* Dressings - a bandage, plasters for small cuts and blisters, larger dressings, gauze dressings, support bandage
* Insect repellant
* Water steriliser
* Eye drops
* Calamine lotion
* Sunscreen
* Disposable needles
* Antiseptic wipes
* Antiseptic cream
* Tweezers
* Small scissors
* Water sterilizing tablets
* Cold and flu remedy
* Lip balm
* Sunburn cream
* Travel sickness pills
* Your own special medical requirements - pills/medication

Remember, prevention is better than cure. So go away healthy and prepared and you will return feeling the benefit of your travels.

A - Z

Airports

Airports can be confusing places. Think hard and ask advice from your travel agent if you are unfamiliar with boarding procedures at airports. Remember to check-in your luggage all the way to your final destination. Keep one bag on flight with you. Keep your boarding pass and passport handy for going through passport control. Most airports now run standard security checks, by electronic and human screening. Do not get upset if you are asked to open your coat and empty your pockets; this is standard procedure. If you wish to enter the duty-free area, bring your boarding pass with you. They will stamp your purchases on this and will note your seat number. A boarding call is not a final boarding call - do not panic, take your time. Toilet facilities in the inner departure lounge should be used as they will be spacious compared to those available on the airplane.

When disembarking, make sure you have your passport and travel documents ready for passport control and decide whether or not you have anything to declare at customs. Only goods bought over and above your duty-free allowance need to be declared (go through red area). Be familiar with your duty-free allowances and stay within them for peace of mind (go through green area). In larger international airports, terminals may be two miles apart, so allow plenty of time for commuting between one terminal and the next.

Air travel

Flying is physically stressful. Modern jet aircraft are artificially pressurised at an altitude pressure of around 1,500 to 2,000m. To be rocketed to this height causes your body stress, so you should do everything you can to minimise that stress, particularly on a long flight.

Loosen your clothing; undo your belt or tight buttons. Take off your shoes, but remember that if your feet swell they may be painful to put on again at the end of the journey. Tilt your seat right back and make yourself comfortable. Ask the stewards for eye blinkers or pillows if necessary. Walk up and down as frequently as possible.

Resist the temptation to eat and drink everything offered. Most people find it best to eat lightly before leaving home, and little or nothing during a flight. Rich, spicy or unusual foods will not agree with you while flying. Alcohol is a bad idea, particularly on a long flight, as it adds to the dehydrating effect of flying. Many women enjoy a drink as a way of relaxing but don't overdo it. Tea and coffee are also diuretics. The best choice of drink is fruit juice or plain water.

Eye gel, moisturising lotion and mouthwash can all help to combat the horrible dehydrated feelings induced by long flights. Also carry a toothbrush and paste to rinse your mouth.

Non-smokers should avoid the smoking areas. Smoking reduces the body's tolerance to altitude. It raises the level of carbon monoxide in the atmosphere, so that non-smokers can also suffer the ill effects if seated close to smokers.

Which ticket?

The world of airline tariffs is incredibly complex and unless you are very experienced you will probably need a good travel agent to guide you through the maze. Bear in mind that cheaper fares are available if:

*you are prepared to accept restrictions on your ticket pertaining to time and date of flights.

*you can let the airline slot you on to a flight which it knows is likely to have empty seats.
*you can book in advance.
*you can fly on a standby basis.

The following are the main types of fares available:

First class

Completely flexible fares, valid for a year. Reservations can be changed to an alternative departure date or another airline. Stopovers are permitted. Mileage deviation of up to 20 per cent is allowed at no extra cost. Perks include better in-flight food, a bigger free baggage allowance, sleeper seats with more leg room, special airport lounges, and general VIP treatment on the ground and in the air.

Business class/Full economy class

A special type of fare, devised with the business traveller in mind. Usually involves a premium of anything from 5 per cent to 20 per cent on the economy fare, and for this extra expenditure the passenger gets enhanced in-flight service and more comfortable seating, along with special facilities like executive lounges and dedicated check-in desks. Generally a flexible fare with the same milage deviation concessions as first class.

Point-to-point economy

Valid only for travel between the two points shown on the ticket. The ticket cannot be used for connecting flights with another airline.

Excursion

Offering a saving of 25 per cent to 30 per cent on the full economy fare, this type of ticket restricts the length of your stay and is usually valid for round trips only. Flights can be changed to an alternative departure date.

Apex/Super apex

The Apex (Advance Purchase Excursion) fare has become the main method of discounting. An Apex ticket must be booked and paid for in advance (usually two weeks to a month), and a minimum stay abroad is usually required. It is quite inflexible: no stopovers are permitted and if you need to change your ticket you will have to pay a fairly hefty cancellation penalty. However, an Apex ticket can cost less than half of normal full fare.

Charter flights

For legal reasons, charter flights are technically classed as package tours, so your fare will probably include basic accommodation. Return dates may not be as flexible. It is not unusual to be restricted to returning only seven or fourteen days after the outward journey.

Round-the-world (RTW) tickets

Many airlines or combinations of airlines offer RTW fares which allow significant reductions. The first part of the journey usually has to be booked two weeks to a month in advance, but after that you book the flights as you go. There are restrictions: you usually have to make a minimum number of stopovers and you must keep moving in the same direction - you are not allowed to 'backtrack'.

Children's fares

A child under two, not occupying a seat and accompanied by an adult, will usually be charged at 10 per cent of the adult's fare. A child under two occupying his or her own seat, or any children aged two to eleven inclusive, pay half the adult fare. Some fares do not offer such reductions. Some Apex fares, for example, offer only a one-third discount for children. Some standby fares offer no reduction at all.

Student fares

Students are entitled to a discount of 25 per cent off the full fare, but such fares are becoming less widely used because other fares like Apex offer better reductions.

Multi-sector tickets

Many of the airlines serving the US offer a multi-sector ticket, allowing you to fly within the US at a very reasonable price, sometimes as little as $30 per flight. Again this is an out-of-season offer, usually between November and April.

Over-booking

Most airlines deliberately over-book their flights because they know there will always be a few 'no shows' - passengers who make a booking and then don't turn up. Occasionally things don't work out as planned and a few confirmed passengers have to be denied boarding. Re-check your return passage booking twelve hours before flight departure.

If you are unlucky enough to be 'bumped', as it is known, you should seek compensation from the airline. European airlines have a scheme whereby you are entitled to a 50 per cent refund of the one-way fare for the sector involved, subject to a ceiling of £150, plus expenses incurred during the delay, including accommodation, if necessary.

Many airlines now invite passengers to volunteer for off-loading from an over-booked flight, offering them compensation in return, while others may simply find themselves dumped. Credit cards are a must for these occasions.

Lost luggage

If the carousal stops going round and your baggage hasn't appeared contact an airline official in the baggage-claim area. Sometimes baggage of non-standard shape cannot be handled on the conveyor belt and is brought separately to the claim area.

If your baggage has gone astray you will have to complete a Property Irregularity Report (PIR) on which you describe it, list its contents and give the address to which you want it forwarded. You should ask the airline for an allowance to enable you to buy basic necessities for an overnight stay.

If your baggage never arrives, make a claim against the airline within twenty-one days.

If your baggage is damaged when you claim it, again report it immediately to an airline official, fill out a PIR form and follow it up with a formal claim against the airline.

Bargain holidays

The best bargains are usually package holidays, and even if a package doesn't appeal to you, these holidays can be such good value that it is worth using them as a base from which to explore the country. You can always go off and book alternative accommodation, using your hotel or apartment as a base. Because they are so cheap you will still have saved money.

Package tour operators make a block booking with a carrier - usually a charter airline or shipping company - and a hotel or apartment for accommodation. Unless they sell every seat before the airplane or ship departs they will be at a loss; empty spaces are wasted money. As departure date approaches, they become increasingly nervous and somewhere between eight weeks and two weeks before departure, they become nervous enough to start selling seats cheaply - which is where you come in for your bargain.

You can only avail of such holidays if:

*You can travel at short notice (usually ten days to four weeks).
*You are happy with departure date, hotel and other conditions under which the holiday is offered. You are unlikely to be offered any choice.

Cancellations are another option. The date and accommodation will be fixed, but you will have to provide the same number of bodies as the original booking.

Outside of package holidays and charter flights, the best bargains offered by airlines are out-of-season Apex (Advance Passenger Excursion) fares. These usually have to be booked at least three weeks in advance, they cannot be 'cashed in' like regular airline tickets, and are subject to restrictions on stopovers or changes of destination (see Air travel page 112). Ask your travel agent.

The rules for bargain holidays are:

*Keep in close contact with at least one, or preferably a few, travel agents. Scour their window announcements weekly and when you find a bargain, be prepared to go.
*Think in terms of out-of-season travel, remembering that the season varies from destination to destination.

Black market

A black market for currency exists where individual dealers operate independently and illegally, offering to change your money and travellers' cheques for a better rate than the one officially offered by government regularised banks and exchange bureaux.

It can seem tempting. Not only are you promised a better deal, you may somehow feel that by changing money on the black market, you are a 'real' traveller, outwitting officialdom and getting one up on everybody else. But beware, black marketeers make their living from currency transactions, and the way they make big profits is by cheating unsuspecting tourists. They know you are not in a position to complain to the police. Operating outside the law, anyway, they are very likely to cheat if they think they can get away with it.

Yes, you might get a slightly better rate. But is it worth the risk? Probably not, unless you are in a country long enough to really know your way around.

It is best to establish a relationship with an independent dealer to whom you can return. A hotelier, for example, cannot afford to lose good hotel business by cheating on money exchange. The proprietor of an exchange premises, no matter how run down or scruffy looking, is likely to offer a safe deal. Check too with other travellers. The money dealer who regularly works your hotel or café, and who has changed money for fellow travellers with whom you can check, is safe enough, and may well offer a better rate than elsewhere.

Countries with a black market problem may insist on a declaration of all money on entry and check this against bank receipts on exit. If you want to take some undeclared money for use in the black market remember that you may be searched and any excess funds will be confiscated. If you are taking the risk, be aware of the consequences.

Camping

A cheap, fun way to go. Start small, buying as little equipment as possible and go away for one night, nearby. You can then progress to a weekend or a longer holiday if you catch the camping bug.

Your equipment is obviously vital to the enjoyment of your trip. Go to a specialist shop and explain your needs to them. A tent should be water and wind proof, have good aerodynamic properties and reasonable headroom. It should be easily erected and should pack up compactly. A good shop will have some tents set up in the showroom for you to crawl around in and lie inside, to get an idea of the space and quality of the tent.

The type of sleeping bag you require will depend on the climates you plan to visit. Hot climates require only a light bag of artificial fibres which can be washed easily; cold-weather camping requires high quality bags. You also need a sleeping pad to lie on as you body weight compresses the filling in any sleeping bag where you lie, and the ground will conduct your body heat away. Again, a good store will advise.

Your pack, which will carry all your equipment, has to fit comfortably, especially if you plan to do any hiking. When you buy a bag, bring it home, and get all your gear together, loading up your pack as though you were going on a hike. Strip off and stand in front of your mirror. Observe the two dimples at the lower end of your back - it is here that the main load should lie.

Adjust your shoulder straps until the load fits squarely over these two dimples. Straps should be wide and padded enough to prevent chaffing, lumpiness or excessive fatigue. Zips should be large and made of plastic to prevent freezing, and covered by flies.

If you plan a walking expedition you will need a properly fitted pair of walking boots. Go to a proper fitter and get advice. Break-in your boots by walking around your neighbourhood in them before taking them on a walking trip.

Other items you might like to consider are a compass, a knife, a whistle, a stove, pots and pans, cutlery and plastic cups and plates.

Camping tips

*Choose your spot early in the day, while it is still bright. If land is marshy look for upland, well drained ground. In the sun a shady spot is essential. The ground has to be flat with as few rocks as possible.
*In hot countries hard, sunbaked ground will not take normal taut pegs. 15cm nails are a good alternative. Remember to bring a claw hammer to drive them in (and pull them out).
*On a long trip take strong thread and sailmaker's needle for repairs.
*Remember in some countries it is illegal to camp outside official sites. Check in advance.
*Smaller sites tend to be less crowded.
*Face your tent door eastwards and you won't have to get up to watch dawn break.

Car hire

Big international companies like Hertz and Avis are expensive - it costs money to provide that easy-to-book, uniform service across different borders and languages - but they have the advantage of being reliable. While a local firm will certainly be cheaper, you shouldn't part with your money until you have satisfied yourself as to their reliability.

*Make sure you are insured.
*Choose the smallest car that suits your needs.
*Avoid milage payments.
*Pre-book if possible.
*Read the small print.

Clothes

Dress is a more important consideration for the independent traveller than for the package tourist who has the safety of numbers on her side.

If you are on the move you should only take one set of garments to wear, one set that needs washing and a few spares for emergencies, along with just one smarter outfit.

The actual clothes depend on where you are going and what you will be doing. Do check on climate, enquiring both about night and day temperatures, rainfall and humidity. It's easy to forget that desert nights can be cold, for example.

Dressing for heat

Choose light, loose clothes made from natural fibres. Avoid nylon and other synthetics which don't let air through. Include a hat or shawl to shade your head. For hot, wet climates you might wish to pack a small umbrella.

Dressing for cold

Wear layers of fabrics which trap air, eg wool, brushed cotton, or down padding rather than single, heavy garments, and cover with an outer layer of waterproof and wind-proof material like gore-tex. A hat is also essential, as a large percentage of body heat is lost from the head.

Dressing for mixed climates

Again, avoid bulky garments and go for the layer approach. A vest, T-shirt, light jumper and jacket can be adapted to the conditions as you find them. Bring a lightweight waterproof and don't forget waterproof shoes.

As a visitor to another country you are a guest and no guest wishes to offend her host. It is polite and sensitive to choose ordinary simple skirts, dresses, jeans and tops to take on your travels. It also means you attract less attention and blend in better with the local scene.

Choose clothes that cover up to the same extent as local ones. For example, where local women wear loose baggy ankle length trousers and tops, choose trousers and a loose shirt. Don't wear shorts away from beaches in areas where local costume covers legs. In Islamic countries, choose clothes with long sleeves, full, longish skirts and wear a headscarf.

Clothes which prove unnecessary or which you no longer need can be posted home. Additional clothing can be purchased along the way, if routes or weather change.

Fear of flying

Air transport is one of the safest methods of travel, but tell that to the woman with aviophobia! The very thought of flying can be enough to bring on a range of symptoms - feelings of panic, sweating, palpitations, depression, sleeplessness, weeping spells, even temporary paralysis.

If you suffer from fear of flying, but you want to, or need to travel by airplane, you will have to learn to cope. One source of fear is simply a lack of knowledge about how aircraft work, knowing which sounds and sensations are usual and expected.

Some airlines conduct courses for what they call the 'white knuckle brigade'. Topics covered include: recorded simulations of the sounds to be expected in flight; group discussions where everybody talks about their fear of flying; relaxation routines which help the flier to cope with the anxiety; statistical information about the safety of flying. Talking to regular travellers may also be reassuring.

A deep rooted phobia is likely to require therapy.

Guidebooks

First of all note the publication date of a guide. An out-of-date guide is almost useless. Bear in mind that research will usually take place six months to a year before the book is published. Then consider the general thrust of the guide. You want the one which most closely matches your interests and travel requirements. The following guide to the guides is very general and deals only with the best known and larger series guides. When choosing a series it can be helpful to skim the guide to your own country. Assuming that others in the series conform to the same standards you should be able to hit on the one that suits you.

AA/Baedeker
Maps and lists of practical information without commentary
American Express Pocket Guides
Good listings and maps and useful cultural surveys
Berlitz Pocket Guides
Compact, tightly written guides offering basic guidance
Blue Guides
Intelligent treatment of culture, architecture and history
Fodor's Guides
Good basic information and opinion on sights and accomm-

odation
Frommel's Dollarwise and X Dollars a Day Guides
Good information on value for money, particularly in relation to hotels, restaurants and nightlife
Insight Guides
Descriptive texts, colour illustration and hard information for the sophisticated traveller
Let's Go Guides
Budget guides to Europe and the US
Lonely Planet Guides
Guide to individual travel in adventurous destinations
Michelin Green Guides
Detailed description of sights, history and art. Star ratings and easy to read layouts
Michelin Red Guides
Hotels and restaurants with information on facilities, prices and famous star ratings
Rough Guides
Subjective opinion and hard facts for the budget traveller

Hitching

Hitch-hiking has its dangers but the experience of many travellers is that it is cheap, interesting and often perfectly safe.

Dangers can be minimised by following a number of basic rules. Don't accept a lift from a driver who seems drunk or suspicious. Follow your instincts. It probably is ill-advised for a woman to hitch alone, but it is up to you and your own judgement of the set-up in the country you are visiting.

Try to look respectable without appearing affluent. Stand where traffic is slow, where you can easily be seen and where a driver can safely stop.

Do remember that all you have done is hitch a lift. You do not have to put up with any behaviour you find uncomfortable. And the golden rule of hitching is: when in doubt, get out.

Having said that, remember that your conduct may well be

the deciding factor as to whether this person picks up hitchers in the future. Be courteous and appreciative.

In some countries, notably in Latin America, you will be required to pay a small fee for a lift.

If you find hitching is making you feel constantly uneasy it's not worth it. Find an alternative method of travel.

Hostels

The International Youth Hostel Federation is the umbrella organisation for youth hostels worldwide. It lays down standards for its members which each national association interprets in the light of its own culture and traditions.

The term youth hostel is now a misnomer as there is no age limit for staying in a hostel. The association which runs the hostel is non profit making, and offers accommodation which is clean and basic, at a very reasonable price.

Most hostels only provide dormitory accommodation, separating the sexes, but more and more are making provision for families. Dormitories can have anything from four to 100 beds along with toilet and washing facilities.

It is usually necessary to pay a small membership fee. Once a member you can stay in any of 5,000 hostels in fifty countries throughout the world.

Most hostels provide facilities to cook your own food. Cheap meals may also be provided.

A full list of the world's hostels, along with regulations and facilities of each, is contained in the International Handbook, available from your local hostel information centre. Volume 1 deals with Europe and the Mediterranean, Volume 2 with the rest of the world. Central and South America is the part of the world least well served with hostels, with just a dozen or so scattered through Argentina, Chile, Colombia, Mexico and Uruguay.

Hotels

A hotel is any establishment offering board and/or lodgings to travellers.

Five-star international hotels have their place. They are comfortable, efficient and consistent in the service they provide. But if you are visiting a strange country to see how it works and what makes it 'tick', staying in an air-conditioned skyscraper totally removed from the lifestyle of local people is not the best way to do it.

A businesswoman travelling may have no option but to pay inflated prices for a convenient location. There may also be the question of prestige. Business contacts (particularly in the US) may not be impressed by a decision to stay in the local doss house. Also if you are travelling with young children you will want to be assured of a certain level of cleanliness.

Always ask to see the room before you commit yourself. Scrutinise the toilets, the sheets, the general look of the room. Think about noise. Think also about safety. What are the other people in the hotel like? Is there a lock on the door? Most hotels are safe. They have to be in order to stay in business.

At the cheaper end of the hotel market, and depending on which country you are visiting, are pensions, guesthouses and bed-and-breakfasts. In some parts of the world the government runs good value accommodation - the ponsadas of Portugal, the paradors of Spain, the old raj houses of India.

Bed-and-breakfast accommodation, particularly the type where a person has decided to let a spare room in order to earn some extra cash, is a great way to meet local people. How you enjoy your stay will depend enormously on the family you stay with but if you don't like them or their establishment you can always move on to the next. Such accommodation is widespread in the UK and Ireland, and is spreading throughout Europe and the US.

In Africa and Asia 'rest-houses' fulfil the same function, and again vary enormously in facilities and service.

Accommodation in family homes - guesthouses, bed-and-breakfasts, pensions - are very suitable for women. Because they are smaller and more personal you feel relaxed and safe. You will probably find the staff protective and concerned for your welfare.

In the Third World, small, cheap hotels often house young men from the surrounding countryside who work in the towns. If you stay in such a place you can be prepared for a lot of staring and attention. You might be better off trying to find out where local women stay, if local women travel.

In some countries like Albania or China, you have no option but to stay in a tourist hotel. These tend to be characterless, dreary places.

Airport hotels are good only for quick stopovers; if you intend spending time in the area avoid these soulless monuments to practicality. As a general rule smaller is cheaper is better, with more character, more local colour, more attention and maybe, just maybe, better service.

Jet-lag

The natural processes of the body are affected by daily cycles called 'circadian' rhythms. These circadian rhythms are governed by the five senses - sight, sound, touch, taste, smell - but also by less obvious senses like the sense of place (more obvious in other species, eg migrating birds) or the sense of time (the body's natural 24-hour cycle which regulates our internal clock.

Flying to a different time cycle, therefore, displaces the senses of time and place, and leads to the physical disorientation we call jet-lag. For the holiday maker jet-lag is an annoying waste of precious holiday time; for the business traveller it can be dangerous, lowering mental and physical efficiency by 20 per cent.

The symptoms of jet-lag are: extreme fatigue which affects concentration and performance, constipation or diarrhoea, insomnia, loss of appetite, headache, impaired vision, loss of balance. Adjustment varies from person to person, but a rough guide-line suggests that twenty-four hours recovery time is required for every two hours time difference. Thus, recovery from a flight to the opposite side of the world could take anything up to a fortnight.

Much attention has been paid to methods of lessening the impact of jet-lag. The following diet, developed by Dr Charles F Ehret of Argonne National Laboratory, Illinois, is said to help, by preparing the body over a three-day period prior to flight.

*Start three days before departure day. On day one, feast: eat heartily with a high-protein breakfast and lunch (meat, eggs, cheese, pulses, nuts) and a high-carbohydrate dinner (pasta, potatoes, starchy vegetables). No caffeinated drinks except between 3.00 and 5.00 pm.
*On day two, fast on light meals of salads, light soups, fruits and juices. Again, no caffeinated drinks except between 3.00 and 5.00 pm.
*On day three, feast again, as above.
*On day four, departure day, fast. If you drink caffeinated beverages, take them in the morning when travelling west, or between 6.00 and 11.00pm when travelling east.
*Break your final fast at your destination breakfast time. Do not drink alcohol on the plane. If your flight is long enough, sleep until the normal breakfast time at your destination, but no later. Wake up and feast on a high-protein breakfast. Stay awake and active. Continue the day's meals according to your destination mealtimes.

While this diet does help, it can be tough going in itself. An alternative idea is just to think yourself into the time-zone to which you are going. In the hours preceding your flight, set your watch to your new time, and don't keep converting back.

Take things easy in the day or two after a long flight, even if you feel okay. Don't tackle any more travelling. Don't make any important decisions.

Jet-lag Tips

*Avoid overnight flight when flying east - causes a short night and exacerbates fatigue.
*By the time you are airborne your watch should be set to arrival time.
*Avoid over-indulgence in alcohol.
*Avoid eating too much.
*Give yourself time to rest after you arrive.

Letters from home

If you are taking a long journey, provide family and friends with a list of addresses along your itinerary, and keep them informed as you go.

Main post offices are the most convenient places to forward mail. Letters should be addressed with your name, the name of the town and country where you wish to receive the letter and the words *poste restante* clearly marked in the right hand corner. When you get to the relevant town, head for the main post office, and ask for the poste restante section.

Make sure your second name is printed in large capitals, so that it is filed correctly and therefore easier to retrieve. You will need to show some form of identification to retrieve mail.

Some embassies will also keep mail for you, but do check first. If you are an American Express card-holder, or if you use their cheques, they will accept mail for you too.

Luggage

What to carry your possessions in depends most of all on how many of the necessities of life you can live without. It also depends on the type of trip you are taking - business or pleasure, short trip or long, hotel or tent accommodation - and on whether you are bringing children with you. Carrying two medium-sized cases is better for your spine than lugging one large one about.

Choose luggage with security in mind. Remember that a bag can be opened, can be slashed with a knife or can be stolen. Have lockable luggage in slash-proof material. Make identifying marks on the luggage. Airline tags can easily be ripped off. Don't put your home address on outgoing bags; it is an invitation to a thief!

Regular air travellers will probably find a conventional suitcase the best option. Buy the best you can afford; cheap luggage does not stand up well to airline handling. Hardsided suitcases with strengthened corners and one handle will best withstand rough treatment.

Remember that some airlines may restrict bag size - check before you pack an out-sized bag. Others will strictly enforce rules about number of pieces allowed.

Backpacks should be left to genuine backpackers, particularly packs with hard, outer frames which are horribly awkward. Softsided rucksacks can be more comfortable, and more easily manipulated into awkward spaces, but your bag is behind you, almost on offer to a passing thief.

Whatever type of luggage you choose, carry a foldable accessory bag as well. Everyone ends up with more luggage at the end of a trip than at the beginning (presents, souvenirs, books).

Maps

Buy guide books at home, but maps at your destination. Topographic maps show the general nature of the country: the lie of the land, forests, marshes, courses of roads and railways, etc. A thematic map focuses instead on one particular aspect of the region such as climate, land use, population distribution, etc.

The best map for a traveller is a tourist map which will also include such features as buildings of historical interest, natural features and tourist attractions, etc. Always check the date on the map; an out-of-date map can be useless.

Menstruation and travel

Menstruation can become irregular when travelling. Don't worry; infrequent periods are not a cause for concern, but a bonus. Not only are you saved bother, if you are not ovulating, you'll also be less vulnerable to pelvic and vaginal infections.

Heavier, or longer lasting, periods are a more wearing response to travel stress. You might want to consider going on the pill. It is also possible to avoid periods altogether while travelling by continuously taking the pill. Ask your doctor for advice.

Take your favourite brand of tampons or towels with you - you may not be able to get them abroad. If you are travelling in areas where toilet facilities are poor do not use the toilet to dispose of used sanitary protection. You are likely to block up the sewage system. If there is no obvious place to leave them in a toilet, take an empty bag with you and dispose in a rubbish bin.

Out-of-season travel

Most people in countries where annual holidays are the norm take their break in June, July or August. The travel trade is geared for this and for their other smaller winter sun or ski rush. For the rest of the year the trade must survive on whatever it can get - any profit is better than none.

For this reason out-of-season travel is cheaper. And there are lots of other advantages too. Your destination is less crowded, less likely to be too hot, the get-rich-quick operators will have left or won't have yet arrived, so you are more likely to come in contact with local people.

If you can go out of season (women with school-going children, for example, may not have the option) it is well worth considering, particularly if you are planning a long haul trip. The savings can be really significant.

Out-of-season tips

*Check details carefully. Is it a resort that closes completely out of season, for example, or a town that retains its own attractions when visitors are not there in their hordes?
*Check out local festival dates. It may be out-of-season to you but fiestas and festivals in effect create a mini-season and you will not get good bargains at this time. You will also find your destination crammed with revellers; whether you consider this a plus or minus you should take it into consideration in advance.
*Will normal amenities be open? Will it matter? Could you manage without a swimming pool, for example? Or could you use another locally?

Overseas projects/Volunteer work

Overseas projects are for those who don't want a holiday. Rather they want to go to another country and do something useful, perhaps research or study, perhaps aid work, perhaps an environmental project.

Finding the right organisation is the main thing and takes time. First isolate the area in which you want to work, then make contact with the agencies who accept volunteers. *The International Directory of Voluntary Work* by David Woodworth, publisher, (Annual, Vacation Work) is an invaluable source of reference to opportunities in voluntary work abroad.

The majority of host countries who welcome volunteers need skilled personnel - doctors, nurses, teachers, engineers. Workcamps will accept cheap, unskilled labour, however, for a short-term period, usually two to four weeks.

If you feel you would like to volunteer for an overseas

project it can help to talk to an ex-volunteer. Having tentatively decided on an organisation ask them to put you in touch with a few returned aides.

Most overseas projects provide board and lodging for their volunteers. Some may also pay your air fare. You will be expected to commit yourself to staying for an extended period, usually a year or two.

Package holidays

There are packages, and packages. Crossing the Atlantic by Concorde to cruise the Caribbean Islands is as much a package as two weeks in an apartment on the Costas. A package holiday today does not just mean two weeks of sea and sand. You can get packaged city breaks, Himalayan treks, multi-centre holidays, and activity holidays/adventure travel is now the fastest-growing area of the travel business.

The great advantage of a package, if you find the one you want, is that it frees you from the hassle of having to organise the details. How you get from A to B, where you will stay, perhaps down even to what you will eat, is all decided in advance and is included in the price of your holiday.

Of course if you are the type who loves booking and arranging, considering it all part of the fun, then a package break is not likely to suit you. But it's foolish to be snobbish about package holidays if one can get you where you want to go cheaply and efficiently.

Read the brochures with a real map in front of you, not the artist's impression in the blurb. Examine the sales pitch for advertising jargon that may be trying to disguise the fact that you are miles from the beach, or whatever. Read a good guidebook about your destination. Only then, if the holiday still appeals to you, should you make a decision.

Poverty

When you encounter people without food or shelter, living in slums or on the streets, ill, disabled and probably begging, the shock is devastating. You may have heard it all before you arrived but nothing prepares you for the pity, compassion and disgust that overwhelms you in the face of abject poverty.

You are likely to feel helpless and hopeless, because there is nothing you can do. You are also likely to feel a natural urge to escape the distress.

Do anything you can to salve your conscience. Perhaps set aside an amount of money to give each day. When it is gone, that's that, no more until tomorrow. You could spend part of your travels working for one of the many local or international aid agencies, or you could make them a financial donation.

On your return if you are in the position to do so you could apply to one of the organisations that sends volunteer workers to the Third World.

Poverty is relative, though. There is a danger that those in western societies judge poverty by the absence of washing machines or TVs, for example. Rather we should define those who have and have not by their own society's standards.

Some societies have a tradition of courtesy and hospitality that travellers find difficult to accept. You might be reluctant to take food, accommodation or help from someone you perceive to be much poorer than yourself. This attitude can be insulting. Accept gracefully what is offered and show appreciation.

In some areas so many travellers give so many presents to local children that they quickly come to see any foreigner as a source of easy pickings. Children beg whenever they think they can get away with it, and many of their parents who hold traditional values would be horrified to see them in action. Requests for sweets or 'one school pen' seldom have anything to do with real need and indiscriminate handouts undermine local traditional hospitality.

Give them time, a look through your binoculars, a chance to

practice another language with a native speaker. Show them pictures of your country, your family and home. Share a song or a game. Such activities create an equal bond, connecting in a way that a quick handout never could.

If you promise to do something when you get home, do it. Far too many travellers promise to send books, photos or presents from their own country and never get round to it.

Sexual harassment/Attack

For most women the prospect of bad roads, irksome border regulations or less-than-luxurious hotels fade to insignificance beside the fear of sexual attack. It is this fear which most often prevents women from getting going in the first place and which most limits their travels as they move.

Yet the number of women who have experienced sexual violence or attack while travelling is a tiny minority. Depending on where you are going, and where you come from, you may be statistically safer away than at home. What causes anxiety is that your instinctive knowledge of what is 'appropriate' behaviour for you and threatening behaviour from men is not so finely tuned in different surroundings.

Besides taking common-sense precautions there is little you can do. Sexual attacks do not happen because a woman does this, or failed to do that, but because some man got it into his head. So exercise a reasonable amount of caution, then forget about it.

Avoid poorly lit, unpopulated streets after dark. Don't be paranoid but be aware of situations and people. Trust your instincts. Never hesitate to act because you are afraid to appear silly. Many women are intuitive and can gain a fast impression of a person's intentions from a single look.

Preserve your anonymity until you choose to give it up. Make up a name/home address if pressed by somebody whose motives you doubt. In hotels keep your key number out of view.

Women who travel frequently don't dither. They use body

language to fend off unwanted attention. They proceed purposefully. Give the impression you know exactly what you are doing. Steer clear of risky situations. And, in the unlikely event of attack, have a strategy planned (see Self-defence below).

Sexual harassment, unlike sexual attack, is all too common, and few women travellers will escape this particular plague. The most usual forms of harassment are unwanted sexual comments, pestering or following, uninvited touching or heavy sexual advances.

There is a danger that to give a list of do's and don'ts for avoiding harassment somehow gives the idea that it is women who create the problem when obviously it isn't. At the same time there is nothing to be gained from drawing unnecessary attention to yourself.

Use local women as a guide to what is and is not acceptable in terms of dress. Plan journeys so that you arrive in daylight, especially if you haven't organised accommodation. Don't respond to remarks or noises - it only provokes more attention. Nip unwanted advances in the bud. If casual conversation turns sexual, snub.

But remember it is probably impossible to completely avoid harassment. The best defence is to assume a confident, assertive manner. Sooner or later you will find that the mask becomes real, that you can ignore the harassment and not let it bother you.

Self-defence

Your aim is to escape, never to fight back. If you see a threatening situation emerging shout at the top of your voice and run like hell. You may find a personal alarm reassuring to carry. If you ever have occasion to use it don't stand there, set the alarm off, throw it at your attacker and run, run, run.

If you are followed, go to a public place like a bar or petrol station as quickly as possible. Ask for help or call the police.

You may also find it reassuring to do a self-defence course.

Again remember that the aim of self-defence is not to immobilise or overcome your attacker, but to teach you methods of winding or surprising him which will allow you a valuable few seconds to make your escape.

Some useful techniques

If you are attacked from behind:

1. Stamp on foot.
2. Bring arm forward, then hit backwards into his stomach as hard as possible.
3. Kick backwards aiming at bony part on front of shin.

If you are attacked from the front:

1. Kick the front of shin, or side of knee. Despite what you see in the movies, and all your best instincts, aiming for the testicles is not a good idea, as he could easily grab your leg and a coat or long jacket would protect him.
2. Grab one of his ears and twist.
3. Straighten fingers of one hand and keeping them rigid jab them hard into his throat.
4. Using two fingers jab his eyes.

If these tactics succeed in surprising your attacker, run away as fast as you can.

Check your local telephone directory for details of a self-defence class near you.

Street names

Americans use the grid system, arranging streets regularly in blocks, so that all the streets running horizontally across the city are numbered from top to bottom, while all the avenues running vertically across the city are numbered from right to left. The combination of the two numbers takes you to within a block of where you want to be, and you always know where you are in relation to anywhere else.

Would that the rest of the world was so organised! Some countries are particularly confusing. For example in Japan most people have to hand out small maps with their address to show the best route from the nearest public transport stops.

All you can do is become proficient at reading maps before you leave home. Buy a good map at your destination. Take time to familiarise yourself with your location. Don't stop to consult a map in the street; find a quiet corner of a shop, for example, or the lobby of a hotel.

'Sustainable' tourism

Tourism today is a mass market, affecting the lives of millions of people world-wide. In 1990 over forty million people were classified as tourists and their spending has made tourism today's fastest growing industry.

The traveller is part of a two-way process, departing but also arriving. And because tourism's raw materials are the lives and environments of other people, its impact reaches further than most other forms of economic activity.

Many people living in tourist destinations are now counting the cost of development that failed to put their interests and rights on a par with those of their visitors. Livelihoods are being lost, religions and cultural traditions debased and natural environments degraded by tourism.

As travellers we can all have an influence on how tourism will affect the areas we visit.

*Check with your travel company about their attitude to sustainable tourism and the local people.
*Respect local customs.
*Buy only local craft goods made from sustainable resources.
*Spend your money on local goods and produce, rather than on international hotels/food/drink. Put as much as you can into

the local economy.
*Respect the local environment. Take only photographs, nothing else. Leave nothing behind you.

Swops

Exchanging your home for that of another woman in a different country can be a good way to save money on accommodation costs. It also means you live in a real residential area, rather than a holiday property and so have more contact with the normal life in the country you are visiting. But there can be problems.

There are a number of organisations which will organise home exchanges. Consult your local telephone directory for addresses and telephone numbers. Some just publish a directory, others visit the property and get details of what each member is looking for, then match you with a suitable other half. This obviously costs more but can save you an enormous amount of hassle and time.

Make sure that all eventualities are discussed and agreed in writing. Leave notes about how to work household appliances, about local transport and anything else you think might be useful. A welcome gift, a bottle of wine perhaps, is a nice gesture. Tell your neighbours what's happening before you go.

At the end of your time in somebody else's home make sure you leave it as you hope they have left yours - clean, tidy and with anything broken replaced or paid for.

Theft

It is known the world over that travellers, even the humblest backpackers, carry cash, passport and camera at a minimum, so no matter where you go, you are vulnerable to theft. Thieves operate in a opportunistic way, taking advantage of situations as they arise. If you are too much trouble they will go away and try somebody else.

Plan ahead and take precautions. Take only what you really need, and leave at home anything to which you are particularly attached. Buy a cheap watch, for example, the sort you could lose without a qualm.

When travelling, never, never pack all your valuables in one place. Carry the cash you need for the day in your bag and pocket and if you must carry other valuables that you don't immediately need - traveller's cheques, passport, the bulk of your cash - keep them under your clothes, preferably in a money belt.

Use pockets at the side or front of your clothes, never the rear. Seal your pockets with velcro and you will hear a thief try to open them. Keep your camera attached in a shoulder bag made of slash-proof material and wear the bag diagonally across your chest with the contents to the front.

If you are going somewhere known for crime carry as little as possible with you and don't wear jewellery or a watch. The less you look like a potential target the less likely you are to have anything stolen.

Timeshare

When you buy a timeshare you are, in effect, renting a holiday home for a period (usually two weeks) for the rest of your life - or for as long as the company stays in business, which, if you are not careful, could be a lot shorter.

You buy your two weeks and all the other fortnights in the year are sold to others, so the standard of accommodation can be more luxurious than what you could afford if you were to buy a property outright.

Timeshare is sold by different methods. When a freehold is purchased your period of time is yours for good, and you may let it, sell it or leave it to your heirs in your will.

It is an expensive option, however the salespeople might present it. Out of a typical timeshare payment, one-third of the money goes on advertising, one-third on profit for the timeshare developer and the remaining third on the property itself.

*Purchase from a well established developer who is a member of an internationally accredited association. Resorts Condominiums International (RCI) is one such association.
*Use a solicitor for the transaction. Have the wording on all documents checked. Pay particular attention to the maintenance contract and to what happens if the builders or the management company get into financial difficulties.
*The property should be well situated and have adequate facilities. If any of the amenities promised do not yet exist, get a commitment in writing from the vendors that they will be completed. If possible set a date.
*What about exchange facilities? Can you trade your holiday for another or are you confined to spending your holidays in the same place for the same two weeks for the rest of your life? If exchange is permitted, what is the fee?
*How many units are available under the scheme? There should be at least ten if the operation is to be viable. Too many makes for a less pleasant environment.
*Wherever possible talk to an existing owner before purchasing.

Trains

Train ... it seems impossible that the same word describes the efficient and expensive gliding machines in parts of Europe and the hard-seated, chicken-infested chuggers of parts of South America.

The big advantage of train over coach travel is that there is space to move around, although in some countries, notably Italy and India, overcrowding prevents this. You have better opportunities to meet fellow passengers on a train, however, and on some trains you can eat and even sleep, enjoying a most restful form of travel.

Consult the Continental Timetable for details of train travel within Europe, and the Thomas Cook Overseas Timetable for rail information throughout the rest of the world. General comfort varies from country to country and from railway to railway. Europe is good, as is the US although American railways are few and far between. South America and Asian trains tend to be less efficient and less comfortable, but much cheaper.

Transport

Most countries offer a range of facilities: airplanes, trains and coaches for long-distance travel; taxis, trains and buses in the cities. There may also be forms of transport unfamiliar to you, eg rickshaws.

The degree of comfort, and the relative price of a means of travel can vary enormously from country to country, largely depending on the extent to which it is subsidised by the government. One of your first tasks on arriving in a country should be to establish what you are dealing with. You may not wish to choose the cheapest method, or always want to travel in luxury, but having the relevant information is important.

Private operators may also provide coaches, shared taxis or mini-buses. In some countries these may not follow a timetable, simply waiting until the vehicle is full before setting off. Similarly they may not have a route with recognised stops, but just head for the nearest town, picking people up and dropping them off wherever they want to go.

Make sure you have plenty of small change for payment. Some systems are incomprehensible. Consult fellow passengers about how you pay, whether it is by ticket machine, tickets bought from a local shop, or payment to the conductor. Your guide book might help. If all else fails consult the tourist office.

The most difficult situation is when you have to negotiate a trip, particularly if you have no idea what the going rate is. Your first trip from the airport or port will probably be overcharged - see if you can find some other tourists to share the ride with you. Thereafter ask around among the locals before you travel. If this isn't possible, let locals pay first and see how much they hand over.

When negotiating, it's best to settle before you travel. That way you can go, or threaten to go, elsewhere. It's much more difficult to argue when the journey is completed. Whatever your means of travel remember:

*Be flexible about your route and examine all possible routes and means of travel to your destination in terms of comfort and price et cetera, before making a final decision.
*Whenever you arrive in a new place, try to find out about transport methods and how far ahead they are booked up.
*Prices rise dramatically whenever your route crosses a border. Usually you can save quite a bit by taking a bus as near to the frontier as you can go, then walk across and continue your journey by local transport in the new country.

Women-only transport facilities

Many countries provide women-only compartments on trains and buses and women-only waiting rooms. On the plus side you are free from male attention and your luggage is safer. You may also find it the best chance you have of communicating with women of the country you are visiting, as they are more likely to be loquacious in the absence of men.

The minus side may be the children. There's nothing worse than screaming babies when you are travelling, particularly if you have left your own at home.

Women's groups

Most large cities of the world have women's organisations or women's centres which are a good way of meeting local women and getting a perspective of what it is like to be a woman in that country.

Another source of contact is ISIS-WICCE, an international women's information and communication service, with a comprehensive resource and documentation centre.

They have a network of 15,000 contacts in over 100 countries, all groups or individuals who are working on issues of concern to women. They also have a unique collection of materials by and about women all over the world - periodicals, pamphlets, unpublished material and a feminist library of 14,000 books.

Further information from:
ISIS - WICCE Switzerland
Case Postale 50 (Cornavin)
CH-1211
Geneva 2
Switzerland.

Another organisation of interest is Women Help Women, interested in promoting international friendship among women by helping female travellers to stay with other members and their families.

Contact:
Women Help Women
c/o Granta
8a Chestnut Ave.
High Wycome
Bucks HP11 1DJ
UK.

Working trips

This section is aimed at those taking a long trip away, either in one country or on a round-the-world trip. Working in a country is the best way of getting inside a foreign culture. It also helps to pay for the expensive business of getting around.

How much you decide to save before you go depends on your attitude to money. Once you are resolved to go, set a realistic target amount to save, estimate how long it will take you to raise the desired amount and do everything you can to stick to the deadline.

If you have very little spare cash do invest in a return ticket.

Getting a job before you go

If you have ever worked for a firm with branches abroad, it may be worth seeing whether there are any such prospects of work in the country you wish to visit. Hospitals, schools and businesses may advertise in your home country, or you could apply to them direct.

Another option is to apply to one of the many organisations or agencies active in the country you wish to visit. Finally, is it possible to make a contact in the country you hope to visit? Some cousin of a friend of your local shopkeeper, perhaps?

If you have special skills which are in demand in a particular country it may be worthwhile advertising your services in the press there.

Getting a job when you get there

Rural areas are generally easier, but it is possible to pick up work in city areas too. Susan Griffith, author of *Working Your Way Around the World* recommends going for the jobs which are 'least appealing such as an orderly in a hospital for the criminally insane, a loo attendant, pylon painter, assistant in a battery chicken farm, selling encyclopaediae'. Obviously it's up to you.

The same criteria for nabbing a job at home apply: dress neatly, show keenness, be polite and willing.

Unofficial jobs carry an element of insecurity - you are not protected by employment legislation and consequently are unlikely to be able to negotiate with the boss. Often work is available to travellers only because the conditions are unacceptable to locals.

While few jobs undertaken by travellers make them rich, even the unskilled can find reasonably paying work. And by working hard and playing little in a country with a high minimum wage like Denmark, for example, you can fund a long period of travel in a country with low living costs.

English speakers have an advantage over others in that English is a highly sought after language in many countries of the world. In Japan, for example, a university graduate of any subject can earn very good money teaching English.

Other good opportunities lie in the tourism industry - hotels and restaurants, pubs and clubs, or, if you have more than one language, working as a courier or guide. If you are fond of sports and children you might enjoy work at that North American institution now sweeping Europe, the sports camp. Ski resorts provide plenty of opportunity for work too - grooming snow, repairing skis, cleaning chalets.

It's best to apply early for work in the tourism business, some months before the summer season begins. The other way to approach it is to go from premises to premises at your destination and hope for the best. If you do get offered work ask for details of your pay and conditions, in writing if possible.

Agriculture is the traditional source of employment for the casual worker. Harvest time is a particularly good time to arrive in a rural area looking for work, but there are many stages in the farming process which call for short-term labour. If you arrive in an area at harvest time ask around whether there are farmers short of help. Word of mouth works wonderfully well in rural areas. Alternatively, you might get a list of farms needing workers from a farm co-operative or local newspaper.

Business and industry also provide employment opportunities - shops, factories and offices can all experience a rush during peak business times, or when regular staff are on holiday. Again, the best approach is to turn up and let people know of your existence. References are obviously an advantage.

Paid work is not the only option open to you. Travellers often volunteer their labour in exchange for accommodation and food. Again as English is so much in demand many locals will offer long-term hospitality to a native speaker of English prepared to give lessons.

Charities and organisations run world-wide projects which require volunteers, although only those committed to the specific project are likely to volunteer, as payment only covers modest living expenses (see Overseas Projects above).

Check-lists

These lists are intended to serve as memory-joggers only. You may not want to bring all the items listed, but you should find it helpful to photocopy the lists of interest to you, ticking off items as you prepare for your trip.

CLOTHES
*Underwear - pants, bras, stockings, socks
*Skirt and dress
*Long sleeved shirt
*T-shirts
*Trousers
*Walking shoes
*Sandals/beach shoes
*Hat
*Jumper and/or cardigan
*Waterproofs
*One piece swim-suit and/or bikini
*Scarf

TOILETRIES AND PERSONAL ITEMS
*Soap
*Sponge/facecloth
*Shampoo
*Conditioner
*Deodorant
*Nailbrush
*Talcum powder
*Scent
*Cleansing/moisturising creams
*Suncare lotions
*Sunglasses

*Toilet paper/tissues
*Tampons/sanitary towels
*Laundry equipment - detergent, laundry brush, line and pegs
*Accessories - belts, jewellery
*Sewing kit
*Travellers' games
*Books
*Diary
*Camera and film
*Universal sink plug
*Presents for locals
*First aid kit
*Writing paper

BACKPACKING

*Rucksack
*Tent
*Sleeping bags
*Survival bags
*Sleeping bag liners
*Sleeping mat
*Mosquito net
*Insect repellant
*Boots
*Torch
*Alarm clock or watch
*Ear plugs
*Towels
*Binoculars
*Padlock
*Personal alarm
*Water container
*Water sterilizers
*Knife and spoon
*Tea towel

SELF-CATERING

*Bed linen
*Cutlery
*Towels
*Plastic undersheet for babies and young children
*Bottle-opener/corkscrew
*Toilet paper
*Portable clothes line and pegs
*Travel iron and adapter plug if necessary
*Ice-cube bags
*Food for first day

CHECK-LIST FOR BABIES AND CHILDREN

Medical (see page 40)

Toys for trip

*Books
*Cards
*Magnetic games
*Travel sized versions of standard games like draughts, chess, scrabble
*Puzzle/colouring books with crayons or coloured pencils
*Personal stereos
*For babies - teethers, mobiles, building beakers, books

Equipment

*Travel cot
*Favourite toy or comforter
*Soother
*Own cup or trainer mug and dish
*Baby sun protection
*Pushchair or baby carrier
*Bottle-feeding equipment, including sterilizing kit
*Bottle-food heater
*Food for fussy eaters

*Nappy changing equipment
*Nightlight
*Travel potty
*Bibs
*Safety harness/walking reins
*Safety equipment for accommodation, eg locks, stair guard
*Battery operated single element water heater for mugs

Further Reading

Bloomsbury Publishing, *Family Travel Handbook*. London: Bloomsbury, updated annually

Davies, Q and Jansz, N, (ed.), *Women Travel - Adventures, Advice and Experience*. London: 1990

Chester, C, *Going Alone - The Woman's Guide to Travel Know-how*. London: Christopher Helm, 1987

Dawood, Dr R, *Travellers Health. How to Stay Healthy Abroad*. London: Oxford University Press, 1990

Elkington, W, *Holiday Money - The Best Money Deals for Travellers*. London: Rosters, 1987

Ferrari, *International Places of Interest to Women*. USA: Ferrari Publications, 1986

Fisher, R, (ed.), *Fodors Railways of the World* New York: David McKay, 1977

GAIA's Guide, International, 1991/1992

Gorman, S, *The Travellers Handbook*. London: Wexas, 1991

Griffith, S, *Work Your Way Around the World. Peterson's Guides.* Oxford: Vacation Work, 1991

Grossman, S, *Have Kids, Will Travel - The Complete Holiday Handbook for Parents*. London: Christopher Helm, 1987

Haslam, D, *Travelling with Children*. London: Macdonald, 1986

Hecker, H, *Travel for the Disabled - A Handbook of Travel Resources and 500 Worldwide Access Guides*. New York: Twin Peaks Press, 1991

MacDonald, F, *The Women's Directory*. London: Bedford Square Press, 1991

Morgan, R, (ed), *Sisterhood is Global*. London: Penguin, 1985

Moss, M, and G, *Handbook for Women Travellers*. London: Piatkus, 1987

New Internationalist, *Women: A World Report*. London: Methuen, 1985

Sereny, G R, *MsAdventures - Worldwide Travelguide for Independent Women*. New York: Chronical Books, 1978

Robinson, J, *Wayward women: A Guide to Women Travellers*. Oxford: OUP, 1990

Russell, M, *The Blessings of a Good Thick Skirt, Women Travellers and Their World*. London: Collins, 1988

Sewell, H, *Volunteer Work*. London: Central Bureau, 1992

Associations

Amnesty International
International pressure group working for the improvement of human rights and the release of political prisoners.
Amnesty International
1 Easton St
London WC1
UK

The Experiment in International Living
Non-profit organisation aiming to promote international understanding through a homestay programme for groups and individuals around the world. Also offers a training and orientation programme for people wishing to experience another culture.
Contact:
Experiment in International Living
Upper Wyche
Malvern
Worcestershire
WR14
UK

International Globetrotters Club
Informal association of travellers from all over the world with an interest in the cultures and peoples of other lands. GC organises overland expedition in Africa, Asia and Northern Scandinavia, as well as exploratory projects and speciality holidays.
Contact:
International Globetrotters Club
ERC Ave.
Lonise 89B - 1050

Brussels
Belgium
International Union for the Conservation of Natural Resources (IUCN)
Co-ordinates the work of various charities working in the field of conservation.
Contact:
IUCN
World Conservation Centre
Ave de Mont Blanc
Ch - 1196 Gland
Switzerland

International Youth Hostel Federation
9 Guessens Road
Welwyn Garden City
Herts AL8 6QW
UK

ISIS - WICCE Switzerland
Case Postale 50 (Cornavin)
CH-1211 Geneva 2
Switzerland

Tourism with Insight
A group of concerned organisations interested in promoting sustainable tourism worldwide.
Contact:
Tourism with Insight
Hadorfer Strasse 9
D-8130 Starnberg
Germany

Vacation Exchange Club
Longest established home exchange agency in the US.
Contact:

PO Box 820
Haleiwa
HI 96712
USA

Vacation Work International
Produces useful publications on work and study abroad.
Contact:
Vacation Work International
9 Park End Street
Oxford OX1 1HJ
UK

Women Help Women
c/o Granta
8a Chestnut Ave.
High Wycome
Bucks HP11 1DJ
UK

Worldwide Home Exchange Club
45 Hans Place
London SW1X 0ZJ
UK

Index

Absence
 from home 21, 30, 43
 from work 21
Accommodation 15, 25, 34
 39, 97, 114, 115, 116, 117,
 124, 125, 129, 133, 135,
 138, 139, 146
 Hostels 124
 Hotels 18, 19, 25, 40
Africa 47-55
Airports 37, 93, 111, 112
Air travel 112-116
 (*see also* Tickets)
Attack
 (*see* Sexual harassment)
Australia
 55, 58-59
 Climate 55
 tips 55, 58-59

Bags/Luggage 20, 35, 40, 115, 118, 129
Bargain holidays 116-117
Black market 59, 78, 117

Camping 53, 58, 65, 78, 86, 18-119
Canada 83-86
Car hire 26, 58, 120
Check-lists 147-150
 Disabled traveller 25
 Money and paperwork 94

Children 32-43, 97, 99,
 before you go 33, 34
 by boat, train, sea 38-39
 destinations 34
 health 42
 needs 40, 41
 on airplanes 36-38
 on the journey 35
 suggested medicines 43
 while on holiday 39
 why bring them? 32
China 59-62
 climate 59
 tips 59-62
Clothes 29, 50, 51, 75, 120
 for cold 121
 for heat 120
 for mixed climates 121
Commonwealth of
 Independent States
 (*see* Former Soviet
  Republics)

Destination 45
Disabled traveller 26
Driving 58, 68, 94
Drugs 82

East Africa 52, 53
 climate 52
 tips 52-53

Europe 62, 65-66
  climate 62
  tips 65-66

Flights
  Charter flights 114
  internal flights 55, 58, 78, 82
Former Soviet Republics 76-79
  climate 76
  tips 76, 78

Guide books 45, 122-123

Health 8, 24, 29, 96-110
  accidents 96, 97
  AIDS 98, 104
  bites & stings 98, 99, 100, 104, 110
  Check-list 109
  Cholera 101, 108
  constipation 100, 107, 127
  contraception 100
  Cystitis 102
  Food Hygiene 102
  heat exhaustion 103
  Hepatitis 103
  Malaria 104
  Pill,The 100, 101, 106, 109, 130
  Pregnant traveller 30, 31
  Vaccination 29, 108
Hitching 50, 54, 82, 123-124

Indian Subcontinent 69-72
  climate 69
  tips 69

Letters from home 128
Luggage (*see* Bags/luggage)

Machismo 79
Macho Culture 58
Maps 129
  Africa 48
  Australia 50
  Canada 84
  China 60
  Europe 63
  Former Soviet Republics 77
  Indian Subcontinent 70
  Middle East 73
  New Zealand 50
  South & Central America 80
  South East Asia 67
  USA 84
Middle East 72-76
  climate 72
  tips 72, 75-76
Money 8, 15, 20, 39, 45, 79, 91-94
  cash 83, 93, 94
  Credit cards 22, 72
  Eurocheques 93
  Travellers cheques 92

New Zealand 55, 58-59
North Africa 47, 50-51
  climate 47, 50
  tips 50-51

Overseas Projects 131

Package holidays 24, 53, 76, 91

Paperwork
  insurance 20, 21, 23, 27, 29, 86, 90, 91, 93
  passport 20, 34, 75, 87-88, 111
  Visas 88, 89
Poverty 25, 69, 133

Self-defence 135-136
  techniques 136
Sexual harassment/Attack 7, 52, 54, 58, 59, 66 134-135
South Africa 53-55
  climate 54
  tips 54-55
South and Central America 79-83
  climate 79
  tips 79, 82
South East Asia 66-68
  climate 66
  tips 68
Swops 138

Tickets
  Apex 114
  Business class 113
  Children's fares 114
  Excursion 113
  First class 113
  Student fares 115
Theft 50, 51, 72, 79, 81, 138-139
Trains 50, 62, 76, 78, 82, 86, 140
Transport (*see* Air Travel, Car hire, Trains)
Travel 10
  Business 17
  Family 22, 33, 88
  Holiday 13, 24, 39
  Out of season 130-131
  Reasons for 9-12
Travel stress 15, 22, 41
Traveller
  disabled 24-28, 133
  (*see also* Check-list)
  pregnant 8, 13, 28-32, 104
  retired 22-23
    special services 23
Travelling
  alone 13, 14, 15, 18, 65, 78
  in groups 16, 17, 25, 34
  with children 11, 32-43
  with companions 8, 14-16

USA 83-86
  climate 83
  tips 83, 86
USSR
  (*see* Former Soviet Republics)

Volunteer work 131

West Africa 51-52
  Climate 51
  tips 51-52
Women's groups 142
Women only 76, 142
Working trip 143-146
  (*see also* Overseas Projects)

Notes